Electronics Simplified

A Practical Approach

If you feel your concepts in electronics
are clear then you can straight away close this book.
This book is written by learners for learners.
This book tries to find real-time applications of concepts,
any electronics student would come across.
The book is not the authority
of the subject but is just the authors' dream of their
world of Electronics

We believe:
"Everything is pure logic with a bit of magic"
But nevertheless
"The magic too has some logic."

Electronics Simplified
A Practical Approach

Where logics are embedded on to your brain

- Microcontoller (PIC16F)
- Applied Electronics
- Robotics

Abhishek Roy
Hardik Nagrecha

ALLIED PUBLISHERS PVT. LTD.

New Delhi • Mumbai • Kolkata • Lucknow • Chennai
Nagpur • Bangalore • Hyderabad • Ahmedabad

ALLIED PUBLISHERS PRIVATE LIMITED

1/13-14 Asaf Ali Road, **New Delhi**–110002
Ph.: 011-23239001 • E-mail: delhi.books@alliedpublishers.com

47/9 Prag Narain Road, Near Kalyan Bhawan, **Lucknow**–226001
Ph.: 0522-2209942 • E-mail: lko.books@alliedpublishers.com

17 Chittaranjan Avenue, **Kolkata**–700072
Ph.: 033-22129618 • E-mail: cal.books@alliedpublishers.com

15 J.N. Heredia Marg, Ballard Estate, **Mumbai**–400001
Ph.: 022-42126969 • E-mail: mumbai.books@alliedpublishers.com

60 Shiv Sunder Apartments (Ground Floor), Central Bazar Road,
Bajaj Nagar, **Nagpur**–440010
Ph.: 0712-2234210 • E-mail: ngp.books@alliedpublishers.com

F-1 Sun House (First Floor), C.G. Road, Navrangpura,
Ellisbridge P.O., **Ahmedabad**–380006
Ph.: 079-26465916 • E-mail: ahmbd.books@alliedpublishers.com

751 Anna Salai, **Chennai**–600002
Ph.: 044-28523938 • E-mail: chennai.books@alliedpublishers.com

5th Main Road, Gandhinagar, **Bangalore**–560009
Ph.: 080-22262081 • E-mail: bngl.books@alliedpublishers.com

3-2-844/6 & 7 Kachiguda Station Road, **Hyderabad**–500027
Ph.: 040-24619079 • E-mail: hyd.books@alliedpublishers.com

Website: www.alliedpublishers.com

ISBN: 978-81-8424-776-3

Published by Sunil Sachdev and printed by Ravi Sachdev at Allied Publishers Pvt. Ltd. (Printing Division), A-104 Mayapuri Phase II, New Delhi-110064

Acknowledgement

Finally our thoughts are in black and white. This book would not have been completed without the support and encouragement of many people.

Encouragement and support from each and every individual at ISTE Management Team 2011–12, headed by Mr Shivam Shashikant was crucial in all our challenging time. A big thanks to the ISTE Tech Team and our fellow friends from Tronics League 2011–12, who all were immensely helpful in publishing this book. Creative splash of Chirag Vaya (vayachirag29@gmail.com) for the graphic design of this book was very handy. Shashank Kumar and Abhinav Koleti were the pillars of our support throughout the endeavor.

Special thanks to the faculties of VIT, *i.e.*, Dr. Amit Mahindrakar, Associate Professor, Environmental Engineering, School of Mechanical and Building Sciences; Dr. Mathew Mithra Noel, Professor and Division Leader, Applied Electronics and Instrumentation, School of Electricals and our technology partner, Mr. Prashanth B.S, Founder and CTO, HPS Electronics Ltd., Bangalore for valuable guidance and motivation from time to time while writing this book.

The vision and leadership of Dr. G. Viswanathan, Chancellor, VIT University and the Team of Professionals at the VIT have created a unique platform for the students to excel in their career, which needs the most honorable mention.

The Editor and the entire Team at the Allied Publishers for their effort in publishing the book is sincerely acknowledged.

Your queries and feedback are welcome at abhishek.roy.vit@gmail.com; hardik_nagrecha2005@yahoo.co.in

Vellore, Tamil Nadu
15th August 2012

Abhishek Roy
Hardik Nagrech

Introduction to the World of Electronics

This book aims at simplifying the complex concepts of electronics and embedded systems to a level that will not only help beginners to comprehend better but also help experts in this field to realize a few vital points in improving the understanding.

This book will also make you realize how basic components in this field such as transistors, capacitors, resistors, etc. can well be developed cost effectively and circuits that serve the purpose quite well. These can be used to implement circuit designs to the extent of basic robots that are conventionally made using microcontrollers. The main point to understand here is that there is always a simpler way to do the task, you just need to find the way to do it.

The book starts with answering the basic questions one may have related to electronics, and once one is clear with the basics, one can move ahead to grasp new concepts and understand errors commonly made in the process.

The book is divided into three sub-categories namely—Basic electronics, Robotics and Microcontrollers and Autonomous Robots. It is suggested to follow the sequence while reading the sections in order to understand the further topics. Each of these topics is covered in depth in a practical way, and helps one to analyze and understand the concepts and to simultaneously test the concepts grasped therein.

The authors have attempted to help you understand the basics and advanced electronics, that could be handy, particularly, for the graduate students to build projects with greater clarity and technicality with higher chances of integrating allied fields right from high school science to advanced robotics. So go on, read the book and realize the benefit! Enjoy reading!

Contents

Acknowledgement .... *v*
Introduction .... *vii*

Part I: Basic Electronics

1. Resistors 3
2. Application Note 5
3. Variable Resistors 6
4. Capacitors 7
5. Bread Board Basics 8
6. Switches and their Various Configurations 12
7. Understanding Logic Gates 13
8. Transistors 16
9. Logic Gates Using Transistors 21
10. JFETS 24
11. Power Supply Design 27
12. Motors 30
13. How does One Choose a Motor? 32
14. Understanding the Torque—Speed Relationships of a DC Motor 33
15. Direction Control of DC Motor 35
16. Speed Control of DC Motor 36
17. Relays 37
18. Timer Circuits—IC555 as Monostable and Astable Multivibrator 39
19. OP-AMPS 41
20. Optoisolators 46
21. Oscillators 47
22. Sensors 48
23. Combinational Logic Design 55
24. Multimeter Basics 60

Part II: Robotics

25. Manual Robot 65
26. Subsystems of a Robot 66
27. Block Diagram of a Manual Robot 71
28. Interface Circuit 73
29. RF Based Wireless Manual Robot 74
30. Block Diagram of RF Based System 75
31. Mobile Controlled (DTMF Based) Wireless Robot 80
32. Block Diagram 82
33. DTMF and Working of the System 83
34. Designing of Some Robots Using Combinational Logic Design 85
35. Designing a Line Follower Robot 88
36. Designing a Light Follower Robot 90

Part III: Microcontrollers and Autonomous Robots

37. Microcontrollers and Autonomous Robots 95
38. Block Diagram of any Microcontroller 96
39. Introduction to PIC Development Board 97
40. Introduction to PIC Microcontroller Series 99
41. Advantages of PIC Architecture 101
42. Switch Interfacing 107
43. Serial Communication 110
44. LCD Interfacing 114
45. ADC Interfacing 117
46. Timers 122
47. Interrupts 124
48. EEPROM 126
49. Proximity Sensor Based Attendance System 129
50. MQ 2 Based Smoke/Alcohol Detector 131
51. LED Brightness Control 134
52. Speed Control of DC Motor 137
53. Temperature Controlled Fan 140
54. Ultrasonic Sensor Based Distance Measurement 145
55. High Speed Multiple Sensor Based Data Acquisition System 148

Part I

Basic Electronics

Resistors

What is a Resistor?

Visually, one thinks it to be a simple component with two leads and an almost cylindrical body, with 4 colored rings on its surface. But there is a lot more to it. It is one of the most elementary devices but, has a key function. It controls the flow of current.

The property resistance is measured in the unit Ohms and these resistors are usually manufactured with a specific resistance, power rating and tolerance.

Now what is Tolerance?

It indicates the extent of deviation of the observed value resistance. Resistances commonly found are carbon Filament Resistors and have a tolerance of about 5–10%. Metal Filament Resistors have very low tolerance and are therefore more accurate.

So how do we exactly find out these values by just looking at them?

The answer lies in the rings printed on the body of the resistor. This system is referred to as colour coding. The first two bands are clubbed as a number, the third band is the multiplier (ten raised to the number) and the last one is the tolerance.

Refer the figure given below:

Figure 1 shows a one kilo ohm resistor with 5% tolerance indicated by the gold band.

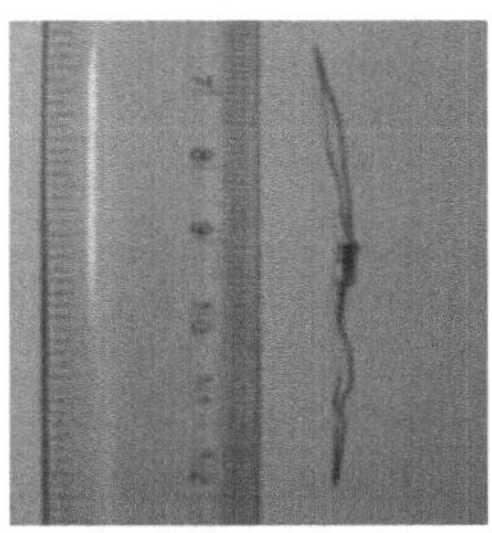

Fig. 1

This component is used in almost all circuits and aids lot of other components. There are two types of resistors w.r.t. their manufacturing/make—metal filament resistors and carbon filament resistors.

Metal filament resistors are comparatively more expensive, more accurate and have lesser tolerance.

They are used in audio signals since they have a very low tolerance of about 1% which ensures that the input signal remains intact after passing through the resistor.

What does the power rating have to do?

One needs to take care of power rating of these resistors. This indicates the maximum power that can be dissipated without damaging the resistor.

The most commonly used ones are ½ and ¼ watt resistors. The corresponding maximum current can be obtained using the formula $P = I^2R$. Thus while using resistors in applications, care must be taken to insure that the current through the resistor is not exceeding the maximum current.

Application Note

How to decide magnitude of series resistor to LED?

One must always remember never to connect an LED directly to the supply, as LEDs have very low internal resistance.

It invites a lot of current through it and it exceeds the rated value and the LEDs burn out.

This is because too much power gets dissipated through the LED.

Thus, we always use a series resistance while connecting an LED to limit the current and preventing it from damaging the LED.

But how does one decide the correct series resistance that one need to place in series with the LED?

You need to know the current rating of the LED to decide the series resistance.

A 5 mm LED has a current rating of 20 mA and hence, considering the supply voltage to be 5 V, we use a series resistor of 220 Ω (approximately equal to $5/(20 \times 10^{-3})$). Similarly, for a 12 V source, we use a 560 Ω resistor.

If you are using a 3 mm LED, the current rating is 10 mA. So you can similarly obtain the required series resistance values.

Variable Resistors

Variable resistors are widely used in a number of electronic applications. Presets, Rheostats, POT (potentiometers), etc. are all various types of variable resistors.

All these devices have a common basic concept, but differ mainly in mechanical structure and operation. Figure 1 shows the symbol of a variable resistance.

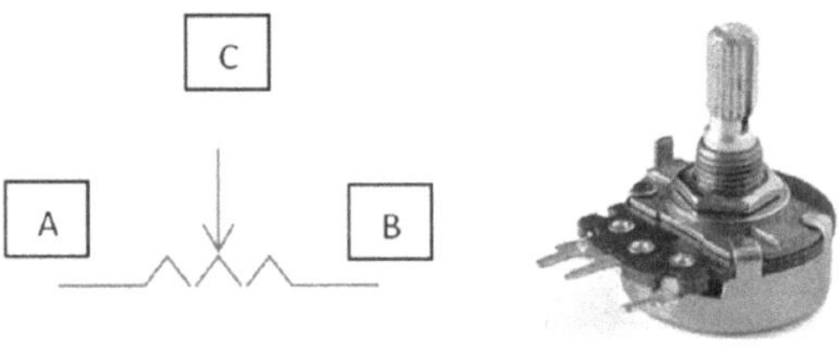

Fig. 1

The terminals *A* and *B* are the fixed terminals while *C* is the variable terminal. To obtain variable resistance, either *A* and *C* are connected or *B* and *C* are connected, or both of these combinations, but make sure you don't connect *A* and *B*. All the variable resistors generally vary the position of the point C with the help of a sliding or rotary mechanism, thus changing the effective length of the resistive element, hence, changing the net resistance.

See Figure 2. It shows a commonly used POT. It has three terminals—one variable and two fixed. The resistance can be changed by turning the screw on top of the blue body. It should be noted that the screw must be rotated till a limit only, or else you will damage the POT. You will hear click sound once you've reached either of the two extreme sides.

Fig. 2

One common application of variable resistor is a potential divider. Refer to the symbol of variable resistors. Now, consider that the terminal *A* is connected to +Vcc and terminal B is connected to ground. Then, variation of the screw will provide an output voltage varying from ground or 0 volts to +Vcc.

We now move on to understand Capacitors.

Capacitors

What is a Capacitor?

This is also a two terminal device with a body. It basically consists of two plates separated by a dielectric. It stores charge.

Capacitors are classified into polar (electrolytic) and non-polar. Non-polar ones can be further classified into mica, paper, disk, etc.

While using a capacitor, two things are needed to be kept in mind—one is the capacitance value and the other is its voltage rating. The capacitance value is generally specified on the electrolytic ones but for the non-polar ones, you need to know the capacitor coding to find out the value (or you should have a multi-meter capable of measuring a capacitance. If you don't know what a multi-meter is, you can read the multi-meter basics in the sections ahead).

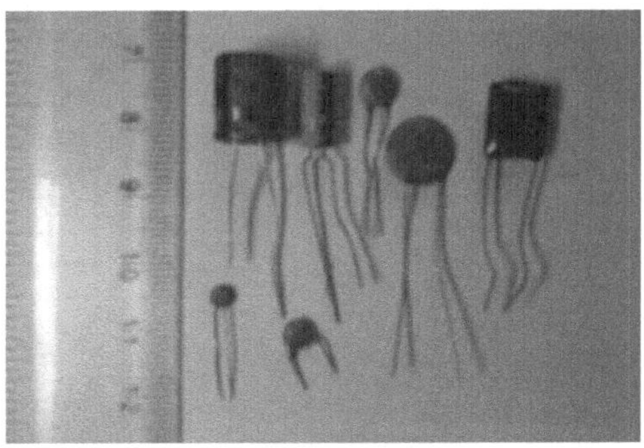

Fig. 1

The figure shows various kinds of capacitors. One should know that these capacitances have various levels of accuracy.

The polystyrene capacitors have the highest levels of accuracy. Depending on the application requirement, the capacitor can be chosen.

Note: In all audio signals it is advisable to use a polystyrene capacitor as coupling capacitor since it has a tolerance of 1% meaning higher accuracies and less noise.

Bread Board Basics

Understanding the Internal Connections

A bread board is one of the most basic tools for developing and testing circuits. It is universally used in schools and colleges because of its simplicity and design, with which one can make connections in no time.

So let us proceed to understand a few basic things about a bread board:

Figure 1 shows the front and back sides of a common bread board. We can see from the back side image as to how the holes of the bread board are internally connected using metal strips.

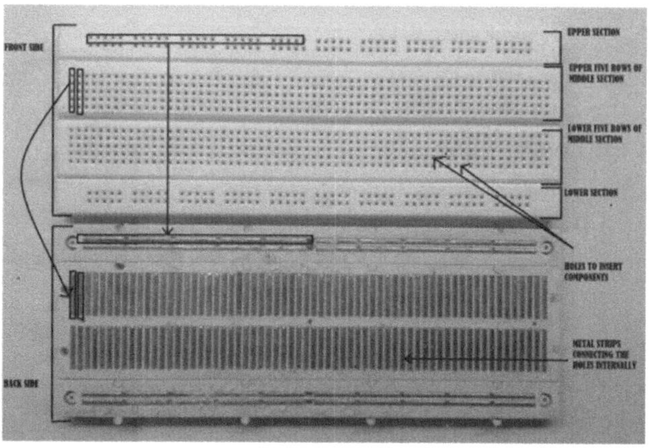

Fig. 1: Bread Board Internal and External View

We first divide the bread board into three parts: the upper two rows, middle ten rows and the bottom two rows. The upper and lower sections consist of 50 columns. The first 25 holes are connected together and the rest 25 are connected together separately.

That gives us a total of 4 groups of 25 holes connected with each other. The figure clearly highlights the connections. (Note that the rows of holes are connected with each other in these sections).

Next, we proceed to understand the middle section of the bread board. It consists of 64 columns and 10 rows. Further we divide this section into upper five rows and the lower five rows.

In the bread board, the 64 columns of middle sections are connected together, the upper five rows and the lower five rows are connected separately, as is evident from the figure. (Note that the holes are connected in columns in the middle section).

How to Connect Various Components and Make Circuits?

Once you have understood how the holes on a bread board are interconnected, it will be very simple for you to proceed to connections. You can insert the leads of whichever component you want into the holes and start interconnecting them then. We first understand the series and parallel connections of resistors.

In Figure 2, you can see that the left set of the resistors are connected in parallel while the right set are connected in series. Can you justify the same?

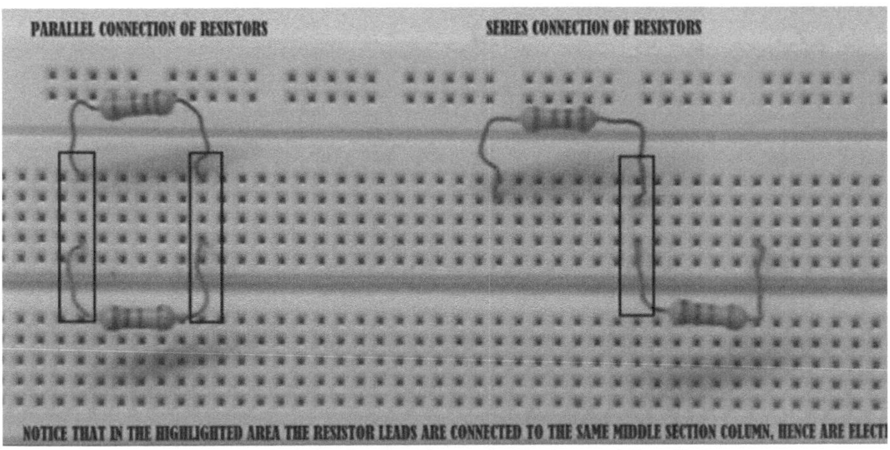

Fig. 2: Connecting Resistors in Series and Parallel

We now proceed on placing the other basic components such as transistors, ICs, diodes, capacitors, etc. (refer Figure 3) and interconnection of these using external single stranded wires.

Note that in some cases we won't be able to connect two components in series directly as we lack space on bread boards. Hence, we use external connections using wires. Figure 4 shows how two resistors are connected in series.

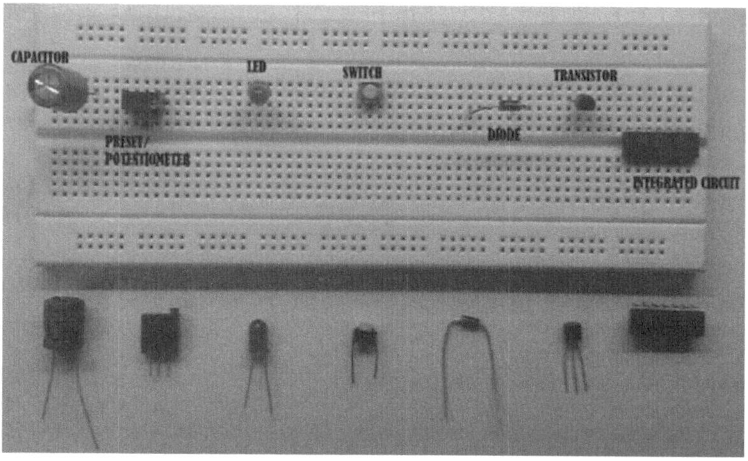

Fig. 3: Connecting Various Components on a Bread Board

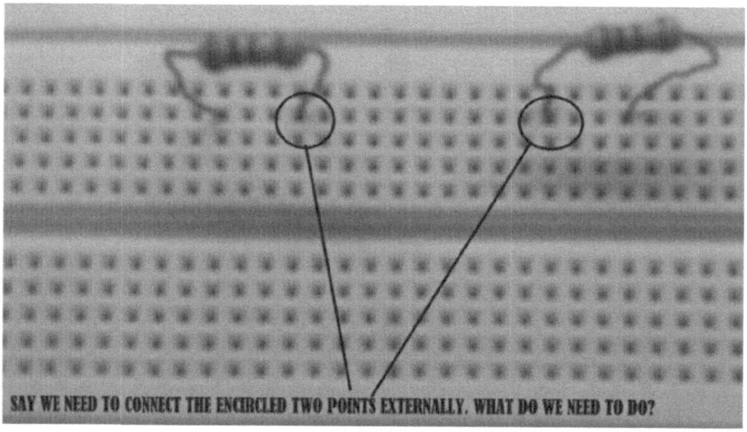

Fig. 4(a)

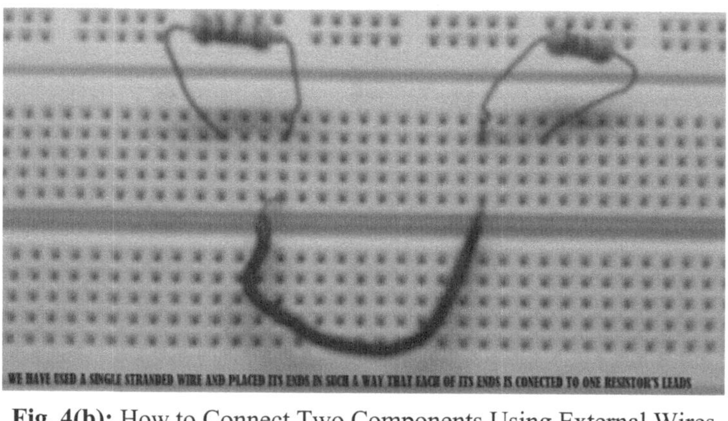

Fig. 4(b): How to Connect Two Components Using External Wires

One more thing to be noted in bread boards is that the upper and lower sections are usually for making connections to a point where a lot of components are connected. Some examples include ground, Vcc, etc. Using these make connections easier to construct as well as check.

Some Common Errors and Suggestions for Connecting Components

1. Firstly, see the correct placement of an IC or Integrated Circuit. Bread boards have a gap in between and are designed to place ICs in the center only. If you place it in any other fashion as shown in Figure 5, you are bound to short two or more pins.

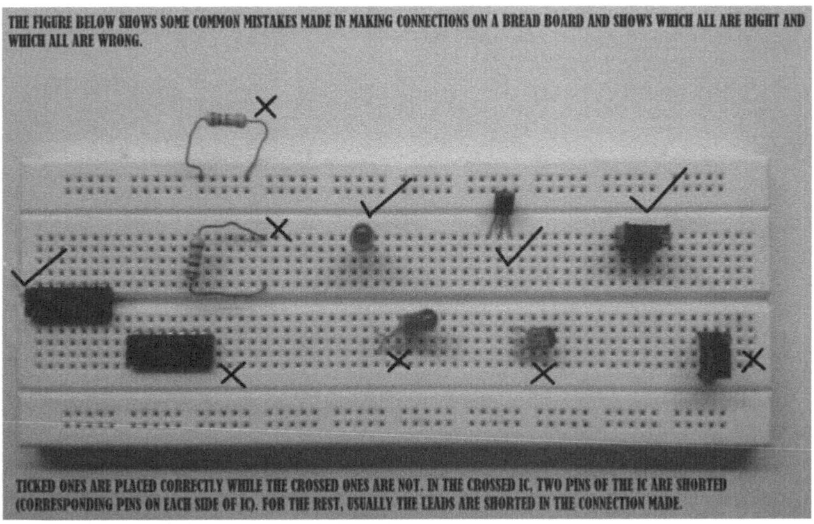

Fig. 5: Some Common Errors while Making the Connections

2. While connecting LEDs, transistors, diodes, etc. make sure you are not shorting the leads of the device. This is the most common error made and if you think you don't understand this point clearly, please re-read the bread board internal connections to understand this point better.
3. Use the upper and lower sections to make connections of ground or Vcc as many devices are connected to these terminals.

Switches and their Various Configurations

A switch is a device which establishes electrical contact between two points in a circuit when actuated, *i.e.* when a certain mechanical or electrical condition is satisfied.

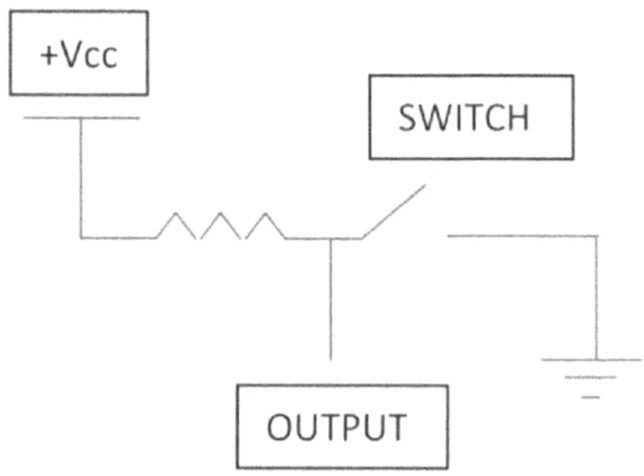

Fig. 1: Active Low Configuration

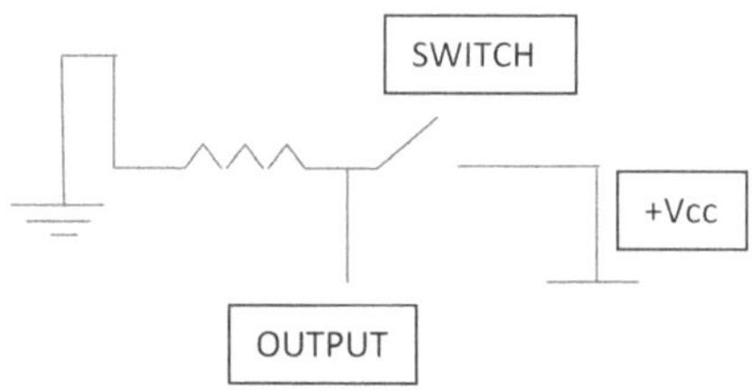

Fig. 2: Active High Configuration

Today, there are various kinds of mechanical, electrical and electro-mechanical switches available in market. Transistors, relays, and various integrated circuits are also available which can be used as switches.

In this section, we are mainly going to discuss two main configurations of a switch—active high and active low.

Figures 1 and 2 show the two configurations. Active high means that when the switch is closed or switched ON, the output is high whereas, the active low configuration would mean that the output would become low when the switch is closed. You can verify the two concepts from the figures above.

See if you are able to understand the output for the switch in ON and OFF/open and closed state.

Understanding Logic Gates

Logic gates are one of the most basic components of various digital circuits used today. You must have learned about these gates in your lower grades, so here's a quick recap of these gates.

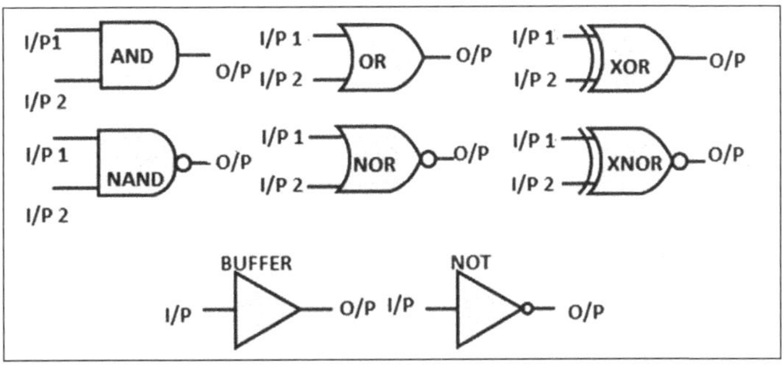

Fig. 1

1. NOT Gate

It is one of the basic gates and is also known as inverter, as the output is the inversion of input. It is usually denoted by $\overline{\text{input}}$ or 'input'.

Truth Table

A	$Y = \bar{A}$
0	1
1	0

2. OR Gate

It is also one of the basic gates and it denotes the sum of Boolean algebra. The output of the gate is high if any one of its inputs is high.

Truth Table

A	B	$Y = A + B$
0	0	0
0	1	1
1	0	1
1	1	1

3. AND Gate

It is another basic gate which indicates multiplication operation of Boolean algebra.

The output of this gate is high only when all of its inputs are high.

Truth Table

A	B	$Y = A \times B$
0	0	0
0	1	0
1	0	0
1	1	1

4. NOR Gate

It is one of the universal gates. It can be regarded as the negation of OR Gate. The output of this gate is high only when all the inputs are low.

Truth Table

A	B	$Y = \overline{A + B}$
0	0	1
0	1	0
1	0	0
1	1	0

5. NAND Gate

It can be regarded as the negation of AND gate. The output of this gate is high if at least one of the inputs is low.

Truth Table

A	B	$Y = \overline{A \times B}$
0	0	1
0	1	1
1	0	1
1	1	0

6. EXOR Gate

It is a gate derived from the basic gates. The output of this gate is high only when odd numbers of inputs are high.

Truth Table

A	B	$Y = A \oplus B$
0	0	0
0	1	1
1	0	1
1	1	0

7. EX-NOR Gate

It is another gate derived from basic gates. It can be regarded as the negation of EX-OR gate.

The output is high only when there is zero or even number of 1s in the inputs.

Truth Table

A	B	$Y = \overline{A \oplus B}$
0	0	1
0	1	0
1	0	0
1	1	1

This concludes the basics of logic gates. You need to be clear with the truth tables of each of these gates in order to understand the further concepts and designs in this book, especially those related to combinational logic design. Review this section again if required.

Transistors

A transistor is essentially a device with three regions namely emitter, collector and base, each consisting of a doped semiconductor.

It can be regarded as two diodes placed back to back, but its characteristics are unique and very useful in developing various circuits such as amplifiers, oscillators, etc.

Transistors are most commonly used as amplifiers or switches in circuits.

They are classified as PNP or NPN (depending on how the three regions of the transistor are doped).

In order to understand the circumstances in which the transistor behaves as an amplifier and as a switch one needs to analyze the graphs of collector current to collect current to collect emitter voltage at different values of base currents.

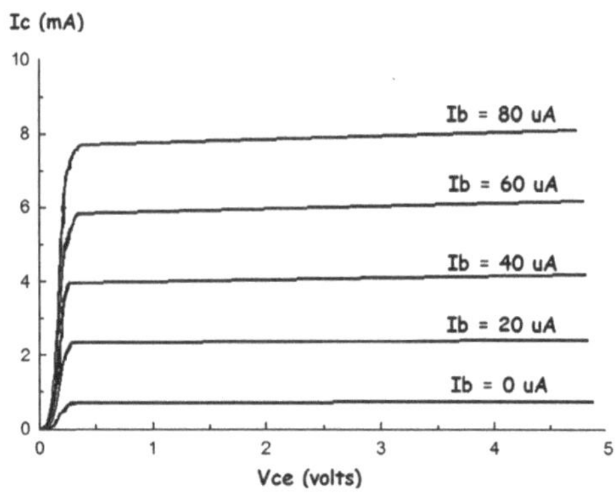

Fig. 1: BJT Response Curves for I_c, I_b, V_{ce}

Observations

As the base current is increased the collector current too increases.

The increase in Collector Emitter voltage due to increase in V_{cc} has no effect on the collector current I_c after V_{ce} crosses around 0.3 V.

What you need to know:

A transistor is essentially a current controlled current source.

The base current I_b is the control signal and the collector current I_c is the element being controlled.

The Q point of a transistor is the operating point of the transistor. It specifies the values of I_c, V_{ce} and I_b of the transistor under concern.

The transistor has three zones of operation namely The saturation zone, the active zone and the cutoff zone.

The saturation zone is the region of the response where V_{ce} is extremely small and collector current is not constant. In order to run a transistor in saturation one must drive the collector current to its saturation.

The active region of the response comprises of the zone above $I_b = 0$ and the region where I_c becomes constant or independent of V_{ce}.

The cutoff region is the zone in the response below $I_b = 0$

There are two bias modes or in simple words the way you can configure a transistor. These are the base bias and the emitter bias.

The base bias is usually run in saturation and cutoff states.

The emitter bias is usually run in the active region.

The gain of a transistor β is not a constant but a function of temperature and Collector current (I_c).

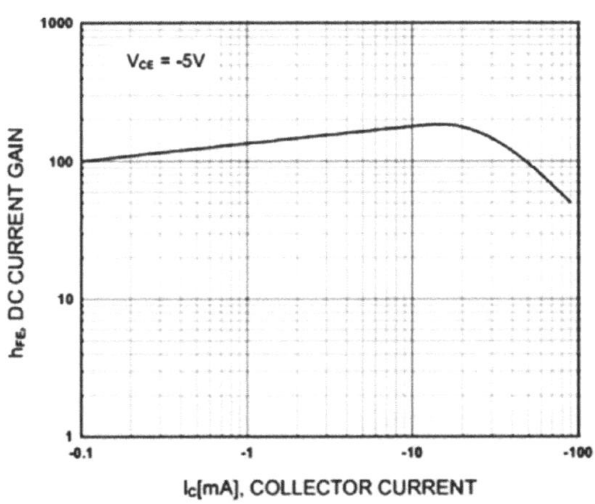

Fig. 2: Variation of Current Gain β vs Collector Current

For switching application it is important the transistor gives two outputs which are well separated so that one of them can be considered as a high/ON state and the other as a low/OFF state. For this it is essential to run the transistor in saturation and cutoff regions.

In amplifiers the most important factor is to ensure that the amplifier has little dependence on the gain of the system. For this we prefer to operate the transistor in the active region.

For switching applications we prefer the base bias and for amplifiers we prefer the emitter bias.

Transistor as a Switch

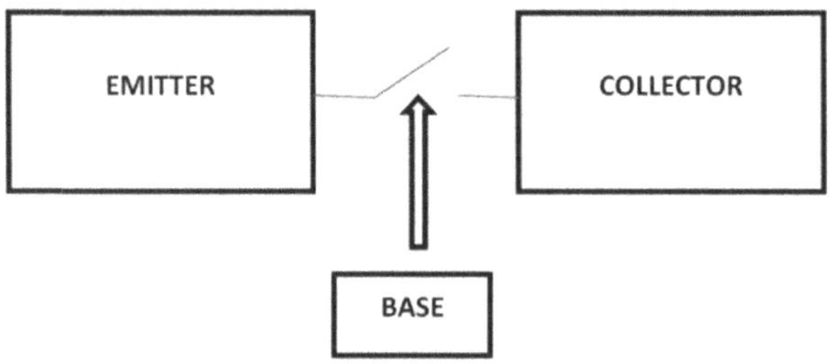

Fig. 3

The above block diagram explains the working of a transistor as a switch.

The main characteristic feature of the base bias is that in this configuration the base current I_b is constant. Even if the transistor is changed or there are changes in operating condition, the base current remains unchanged.

Understanding the Base Bias

By writing KVL equations

Now in order to operate the transistor as a switch it is essential that V_{ce} is very small so that during ON state the collector voltage is approximately zero.

For V_{ce} to be very small it is essential that we set the operating point at saturation.

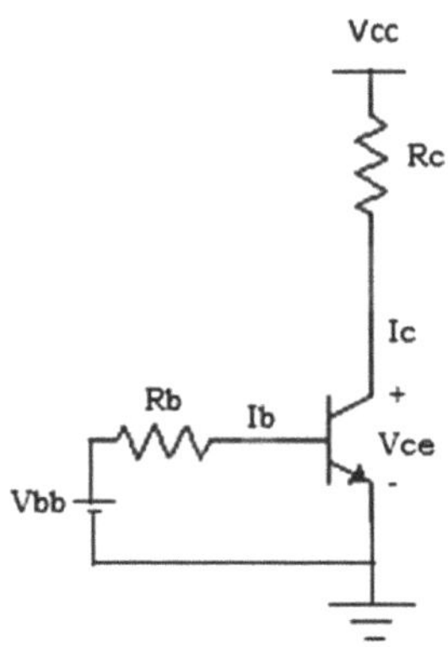

$I_b = (V_{bb} - V_{be})/R_b$

$I_c = (V_{cc} - V_{ce})/R_c$

$I_c = \beta \times I_b$

Fig. 4

But sometimes in transistors there is a huge change in gain from one piece to another, so in order to overcome these shortcomings we set the Q point of our system at Hard Saturation so that all other surrounding factors cannot drive the Q point out of saturation.

How to attain Operating Point in Hard Saturation?

Follow the following approach to find the design parameters:

I_c saturation = V_{cc}/R_c

Assume $\beta = 10$

Calculate I_b from the relation

$$I_c = \beta \times I_b$$

Now using this value of I_b calculate R_b from the relation

$$I_b = (V_{bb} - V_{be})/R_b$$

A circuit designed with the following parameter shows excellent switching characteristics. The basic idea behind hard saturation is to supply the base of the transistor with more base current than required to drive the transistor into saturation. As a result even if β fluctuates the operating point still remains in the saturation zone.

Transistor as an Amplifier

Whenever you want to use a transistor as an amplifier, derive the configuration from the emitter bias.

An amplifier must be designed such that the operating point is immune to changes in β.

Understanding the Emitter Bias

The essential difference between the base bias and the emitter bias is the absence of the base resistor in the emitter bias. This one small change makes all the difference.

The operating point of the new system is rock solid and is immune to changes in β.

Nowhere in the above analysis did we use the current gain β in deriving the relations in the emitter bias.

By removing the base resistor we set up a fixed base current (*i.e.,* neglecting variation of Ib with Vbe) and the emitter voltage is constant since the base voltage minus 0.7 appears at the emitter, giving rise to a fixed emitter current.

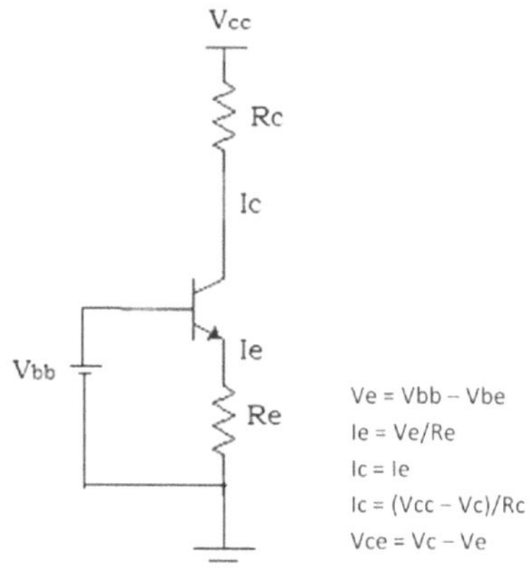

Fig. 5

Types of Transistors

The figure shows some commonly used transistor packaging. Depending on the current rating, the packaging varies. Power transistors have high current rating and are used where high current exists.

You might recollect from the previous sections how logic gates work. We now proceed to develop the basic logic gates using transistors.

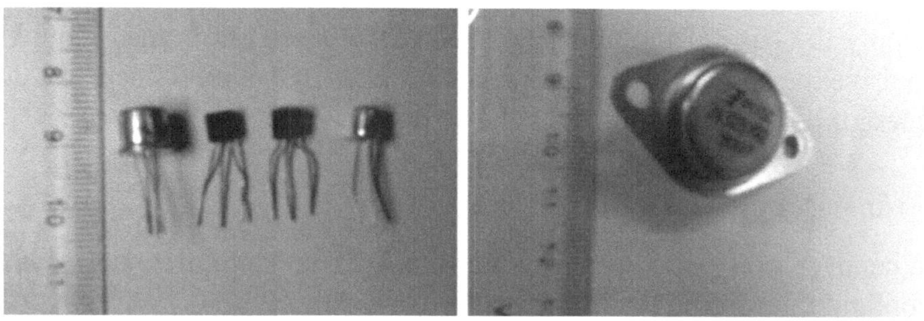

Fig. 6

Note: A transistor is normally supplied with V_{cc} at the collector terminal. What would happen if the base-emitter voltage is greater than 0.7 volts, but there is no V_{cc} (Assuming an LED is connected in the base emitter loop)? The answer would be that the LED would glow dimly because of the small current flowing through the base-emitter loop.

Logic Gates Using Transistors

Transistors can be used to develop all the logic gates used today. Their low cost and easy operation allows us to easily develop basic circuits with loads of applications.

We generally use transistors when we need to use less than 4 logic gates in the circuit so that we don't waste money in buying an IC and wasting one or more of its gates.

In this section, we will learn the basics of developing logic gates using transistors.

We start with understanding the most basic use of transistor, as a switch. We know that if the Base-Emitter Voltage is equal to 0.7 V, the transistor gets activated/is in ON state, and current flows through it, making the emitter current almost equal to the base current.

If the input is grounded, is 0 V, then the base-emitter voltage is less than 0.7 V, and hence, the transistor is in OFF state. No current flows through it in the OFF state.

Once the concept of transistor as a switch is understood, you can easily use it to develop a NOT gate.

Figure 1 shows the NOT gate implementation.

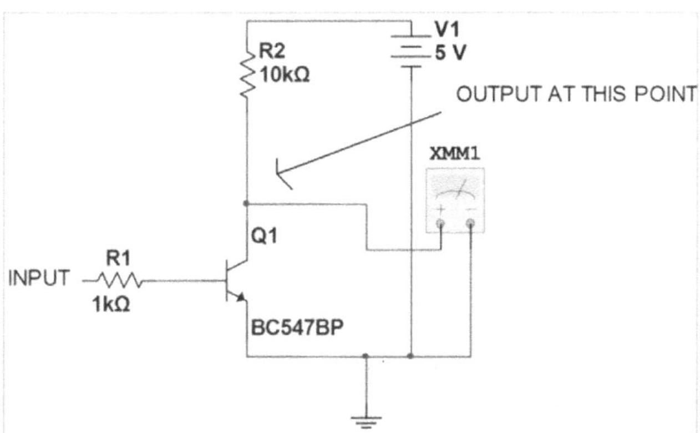

Fig. 1: NOT Gate Using Transistors

Truth Table

Input	*Output*
1	0
0	1

We now proceed to verify the truth table of NOT gate.

Consider the case when the input is low (0). As the input is low, the base-emitter voltage of transistor is less than 0.7 V.

Hence, the transistor acts as open circuit and allows no current to flow through it. Thus at the collector of transistor Q1, where we take the output, is high (1) or 5 V.

Case (1) is confirmed. Next, consider Case (2). Here, the input is high or 5 V, as a result of which the base-emitter voltage is 0.7 V. The transistor is in ON state, it conducts, thus grounding the output. We get a low (0) output.

Hence, Case (2) is also confirmed. Thus, the aforementioned Transistor circuit perfectly acts as a NOT gate.

We now move on to understanding AND gate. Refer Figure 2.

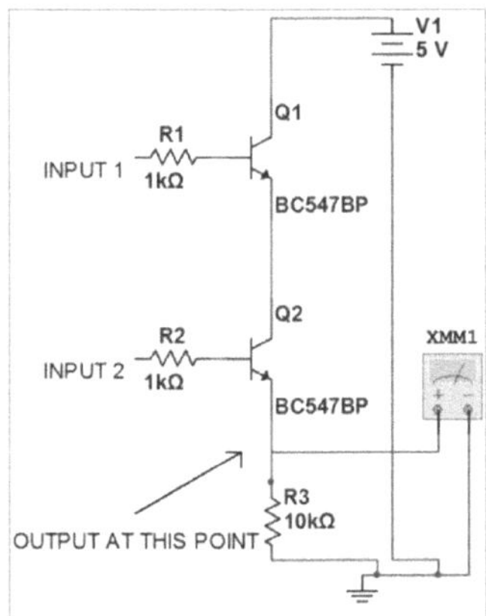

Fig. 2: AND Gate Using Transistors

Truth Table

Input 1	*Input 2*	*Output*
0	0	0
0	1	0
1	0	0
1	1	1

With a similar approach, we can understand how the following implementation of AND gate works. Until and unless both the inputs are high, the output remains low, thus confirming the Truth Table.

Next, we try to understand OR gate implementation. Refer Figure 3.

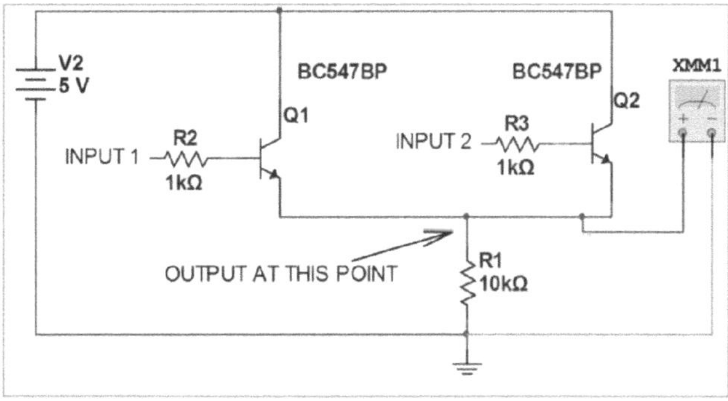

Fig. 3: Or Gate Using Transistors

Truth Table

Input 1	*Input 2*	*Output*
0	0	0
0	1	1
1	0	1
1	1	1

From the figure, we can understand that if both the inputs are low, then the output is low. Else if any one or both of the inputs are high, then the output is high. This confirms that the implementation is correct.

Note: We can similarly implement the Universal Gates, *i.e.* NOR and NAND gates. For the NOR gate, in the OR gate circuit, remove the 10kΩ resistor (placed between common emitter junction and ground) and place it between the common collector junction and V_{cc}. Take the output from the common collector terminal.

Similarly, for the NAND gate, in the AND gate circuit, remove the 10 kΩ resistor and place it between V_{cc} and collector of the upper transistor named Q1. Next, take the output from the collector of Q1.

Thus, using the universal gates or basic gates, we can develop any combinational logic design we want very easily.

JFETS

The JFET is one of the sub categories of the field effect transistors. Now it is natural for a question to arise as to why JFETs? Can't bipolar junction transistors do the job?

The basic difference between a Bipolar Junction transistor and a JFET is that in a BJT there are two charge carriers namely the minority charge carriers and the majority charge carriers.

Whereas in a JFET you only have majority charge carriers, as a result higher switching frequencies can be attained as no additional stored charge has to be removed from the junction.

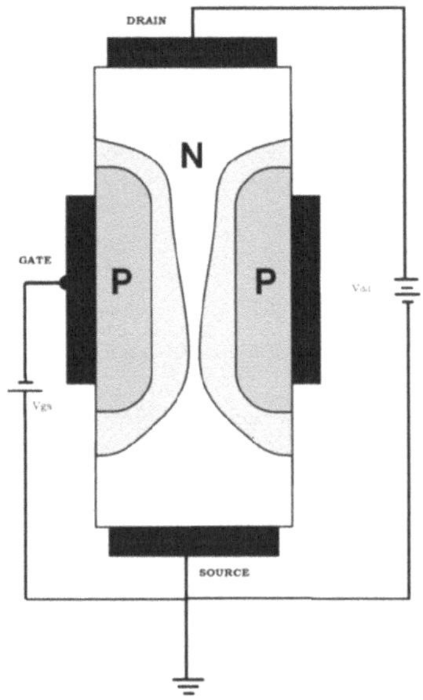

Fig. 1

Unlike the BJT the JFET is a voltage controlled current source.

The JFET has three terminals—the Gate, Source and Drain.

The gate usually consists of a P type material that sandwiches an N type material.

The gate voltage controls the thickness of this channel.

When the channel is thick the current flow is reduced.

The Gate to source voltage essentially controls the thickness of the P layers.

The Gate to source is always reverse biased and the Drain to Source is always forward biased.

The JFET has very high input impedance. This is an added advantage when compared to BJT.

The JFET has two zones of operation namely the ohmic region and the active region.

In the ohmic region the JFET acts like a resistor of fixed value.

In the Active region it acts like a voltage controlled current source.

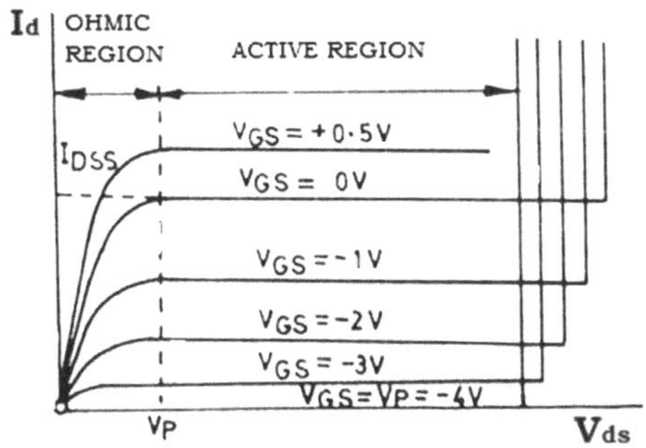

Fig. 2

One observation that we can make from this plot is that the more negative the Gate to source voltage is the lesser is the drain current.

Facts You Need to Know

When Gate to source voltage equals zero, maximum current flows through the JFET. This is also referred to as the shorted gate condition.

The shorted gate condition is essential for the active region.

V_p is called the pinch off voltage and is defined as the voltage above which the JFET enters the active region or the region where the drain current remains almost constant with increasing gate to source voltage.

I_{dss} stands for the drain current with a shorted gate. It is the maximum drain current a JFET can produce.

In the Ohmic region the resistance of the JFET is approximately equal to

$$R_{ds} = V_p/I_{dss}$$

In the plot of Id vs V_{ds} observe that at $V_{gs} = -4$ V the drain current reduces to 0. We call this value of V_{gs} as V_{gs} cutoff. $V_{gs}(\text{off}) = V_p$

FET As a Switch

In the ohmic region the JFET is equivalent to a resistance.

Writing KVL equations one can say

$V_d = V_{dd} - I_d . R_d$

$I_d = (V_{dd} - V_{ds})/R_d$

For saturation,

$I_d \text{ sat} = V_{dd}/R_d$

$I_d \text{ sat} << I_{dss}$

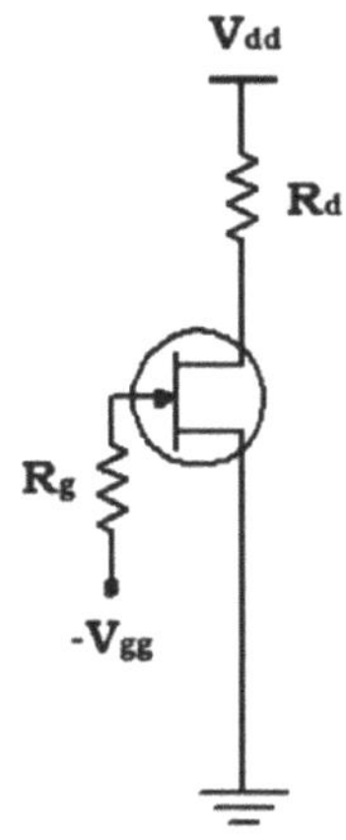

Fig. 3

Under the above conditions the JFET operates in the ohmic region and the entire circuit can be replaced by a potential divider consisting of Rd and R_{ds}

Where R_{ds} is give by the expression:

$R_{ds} = V_p/I_{dss}$

To run the JFET in hard saturation always try to keep I_{dss}: I_d sat = 10:1

JFET find large application in circuits like source followers, choppers, automatic gain control, etc.

The input impedance of JFETs is of the order of several mega ohms. This is another unique feature.

Power Supply Design

Here we discuss how to design a regulated power supply giving 5 V output from 220 V AC line. A regulated power supply is one that can handle fluctuation in both the input and output circuit and maintain a constant output.

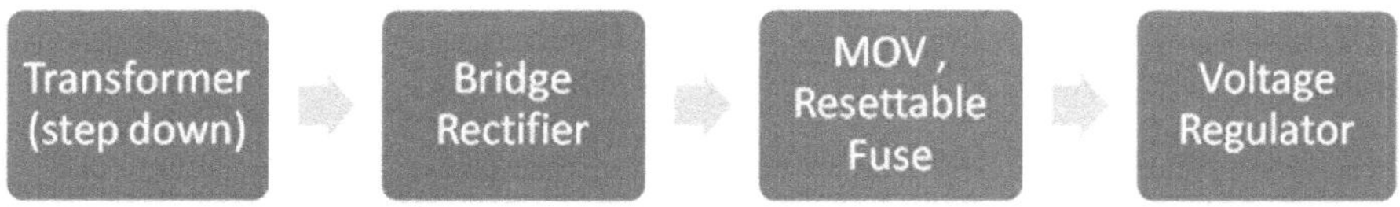

The diagram depicts the sequence in which the various steps take place.

Let us take a look at the role of each of them.

The 220 V AC (Figure 1) line cannot be directly fed to a bridge rectifier as it would require a bridge rectifier having a very large power rating. Hence we step down the voltage to around 12 V which is a general standard in most electronic circuits. Another blunder which students normally do is, they believe if they connect a 100 V DC source to the input of the transformer the output will be a stepped down DC voltage. It must be understood and remembered that the Transformer works only with Alternating currents.

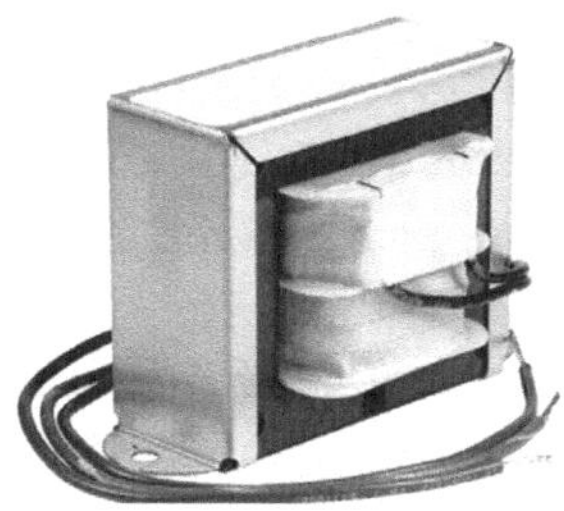

Fig. 1

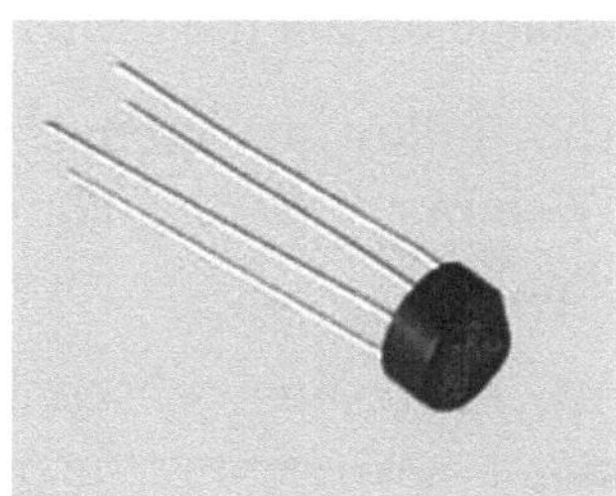

Fig. 2

The output of the Step down transformer is 12 V AC (Figure 2). Now we cannot directly feed it to the voltage regulator as the voltage regulator requires DC input. Hence we use the Bridge rectifier which is essentially a full wave rectifier converting 12 V AC to 12 V DC. The bridge rectifiers come with different current ratings and an appropriate one must be selected.

The MOV or the Metal Oxide Varistor (Figure 3) is often put in parallel to the output of the Bridge rectifier. The main idea behind this is that the MOV allows a voltage equal to that of its rated voltage across its two ends. If the voltage across its two ends goes beyond the threshold then it changes its resistance to draw a higher current through it till the normal conditions are restored. So it is essentially a protective element used for voltage protection of the circuit.

Fig. 3

The resettable fuse has a current rating (Figure 4) and is generally placed in series to one of the output lines of the bridge rectifier. It prevents currents higher than the rated current through it by acting as an open circuit under such conditions and as soon as normal conditions are restored it becomes a short circuit.

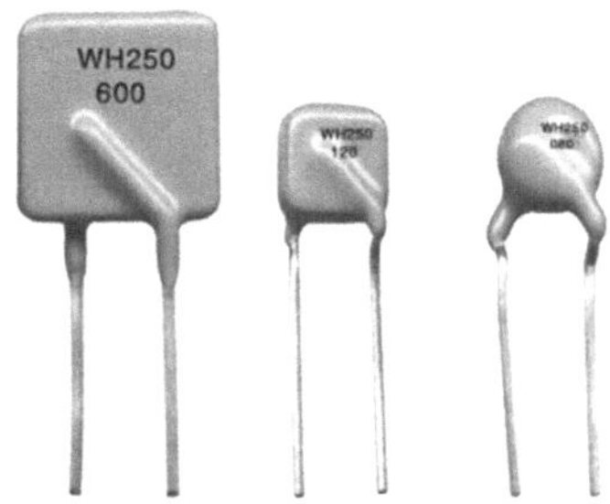

Fig. 4

The 78XX series (Figure 5) of IC are used as voltage regulators. They are all three terminal ICs which have an Input, Ground and Output. The Input must be ideally 2–3 V greater than the output voltage or the IC fails to work. The ICs produce a constant voltage at the output and are ideal for electronic applications. For variable voltage output the LM317 can be used.

Fig. 5

The following circuit diagram (Figure 6) shows how the components must be connected.

An interesting observation which one might make here is the presence of several capacitors.

It is important to understand the role of these capacitors. These capacitors perform regulatory action and are extremely useful when you have a load which draws a large amount of current.

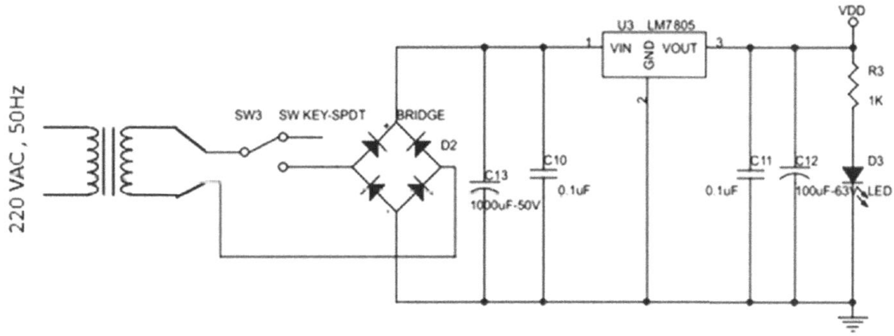

Fig.6

To decide the values one can assume the following approximation. For every ampere of current place a 1000 uF capacitor in parallel to the system. One can place a few 0.1 uF capacitors across the length of the various points to which the supply wires extend.

Motors

Motors are **rated by** their voltage, current draw, output speed, and torque. **Small DC motors** are designed for operation in the 3–6 volt range.

Larger DC motors designed for heavy-duty applications usually require 12 to 24 volts. Often, the higher the voltage, the more powerful is the motor (Figure 1).

Fig. 1

Most motors can be run at **operating voltages higher or lower** than the specified rating. **Lower voltages** reduce torque and speed; **higher voltages** increase torque and speed.

We now discuss two commonly used motors namely, stepper and servo motor.

Stepper Motors

They are used in applications requiring **precise positioning.**

Fig. 2

On providing **current to the coils in a predefined sequence they rotate** (Figure 2).

To **rotate in other direction** the sequence should be reversed.

To **rotate faster,** the sequence needs to be executed faster.

If one **simply energizes one coil the motor** will just jump to that position and stay there **resisting change**. The stepper motor will **generate a lot of heat at standstill**.

As you can understand, this **control can be quite complicated**. So it requires a dedicated controller.

Servo Motors

Servo motors are motors **integrated with a control system** that enables **precise positioning**. So, usually, servo motors **can rotate to a given angle** (Figure 3).

Fig. 3

In a DC motor, the speed and current drawn is affected by the load, whereas in servos the torque, hence current drawn tends to remain constant.

The control **circuitry** to perform operation of a servo motor is much **more complex** than the circuitry that controls a stepper motor.

Because a control circuit is already integrated in the servo motor, usually, you just have to connect it to the power source and provide a control signal. Control signal is a square wave. **Servo motors can be controlled using pulse width modulation.**

How does One Choose a Motor?

There are literally dozens of types of electric motors but for amateur robotics, the choice comes down to these three:

1. In a continuous DC motor, application of power causes the shaft to rotate continually. The shaft stops only when the power is removed or if the motor is stalled because it can no longer drive the load attached to it.
2. In a stepping motor, applying power causes the shaft to rotate a few degrees, and then stop. Continuous rotation of the shaft requires that the power be pulsed to the motor.
3. A special "subset" of continuous motors is the servo motor, which in typical cases combines a continuous DC motor with a "feedback loop" to ensure accurate positioning.

Understanding the Torque—Speed Relationships of a DC Motor

The current a motor draws depends on the torque the motor produces. Because of the interrelationship of torque, speed, current, and voltage, the constant current operation of a DC motor produces constant output torque regardless of speed.

Given a constant load (*i.e.,* torque) the speed of a motor is solely dependent on the voltage applied to the motor. For DC motors operated at a constant voltage, the speed and torque produced are inversely related (the higher the torque, the lower the speed of the motor).

We now mathematically understand the aforementioned relationships:

$$T = r \times F$$

$$P = I \times V = T \times \omega$$

Where,

'P' is the power, 'I' is the current, 'v' is the voltage, 'T' is the torque, 'ω' is the angular velocity, and 'r' is the radius.

Since the load is constant, the torque will be constant. By load on motor in a robot, one must understand it is the frictional force exerted by the

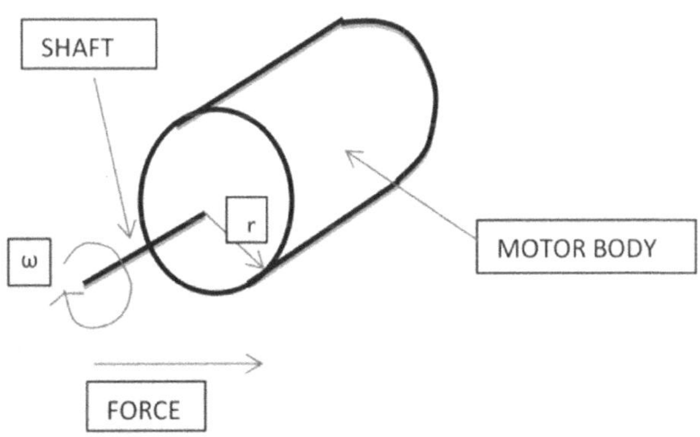

Fig. 1

ground on the wheels, which are in turn connected to the shaft of the motor.

Next, we consider that we apply a certain voltage battery, supplying certain current. Now the value of ω will depend on the battery used (torque being constant and power requirement known). If we use a 6 volts battery for a motor rated at 12 volts, the ω value will be half of maximum.

Direction Control of DC Motor

Consider A and B to be the two poles of the DC motor. Now, if one is kept positive and the other at zero volts, the motor will rotate in a particular direction.

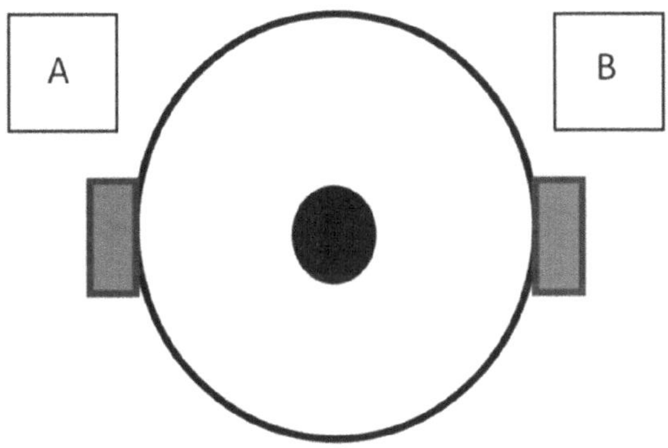

Fig. 1

Say, if A is kept at positive potential and B at zero, the motor will rotate in clockwise direction. If the polarity is reversed, the direction will become anticlockwise.

You can take a normal DC motor and try giving voltages to both the terminals, and see the motion of the shaft for yourselves. Note that if both the terminals are kept at the same voltage level, then there will be no motion of the motor.

Speed Control of DC Motor

Speed of any DC motor may be controlled by varying its input voltage. Lower the input voltage, lesser would be its speed. The armature current drawn by a motor increases with the load it drives.

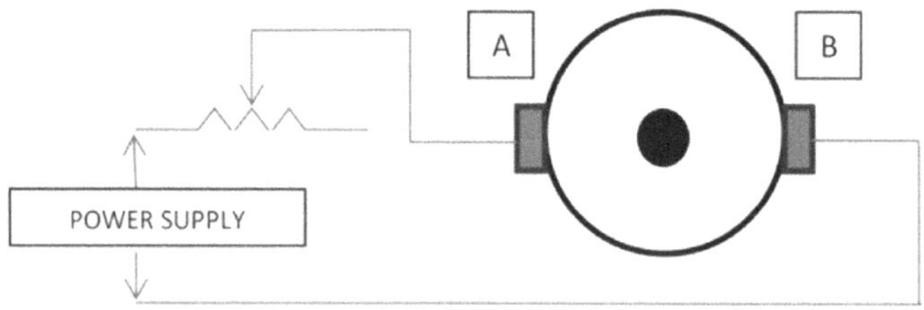

Fig. 1

The above Figure 1 indicates the simplest way to modify the speed of a DC motor. The rheostat or the variable resistor varies the potential drop across it, hence, changing the voltage received by the motor, hence, varying the speed. You can also understand that the resistance variation is varying the current flowing through the motor, hence, varying its speed. Another way to modify the speed of a motor is using a PWM pulse as source. The T_{on} and T_{off} times, we can control the average speed. The pulse causes the motor to switch ON and OFF. If the ON time is greater than the OFF time the motor rotates faster, and vice-versa. Generation of PWM (Pulse Width Modulation) is discussed in the section Microcontrollers.

Relays

Relays are electromechanical devices that are used generally to control high voltage devices using low voltage control signals. It is a switching device which has two inputs and three output terminals. A simple diagram of a relay is shown Figure 1.

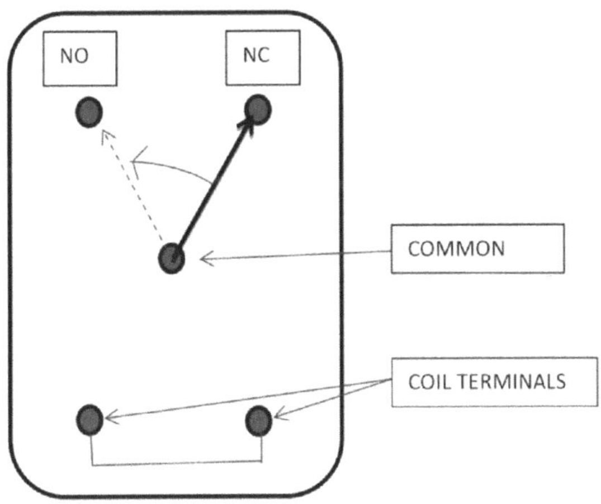

Fig. 1

It is basically a coil of wire wound around a soft iron core. Current flowing through the coil turns the iron into a magnet which attracts a hinged iron lever which, in turn, closes or opens a set of contacts.

The relay coil is the input terminal and there are two outputs—one being normally closed the other being normally open. The third is the common point of contact which usually connects to the device .

The basic principle of operation is that when no coil voltage is applied the common point and normally closed point are shorted on the output side. Upon application of required coil voltage the normally open terminal and the common point short.

The photograph on the next page (Figure 2) shows a sugar cube relay commonly used in various circuits involving interfacing and switching of AC devices using DC signals.

The input required to actuate the relay is 12 volts, and the maximum output it can handle is – AC (7A, 250 volts, 50 to 60 Hz), DC (12A, 120 volts).

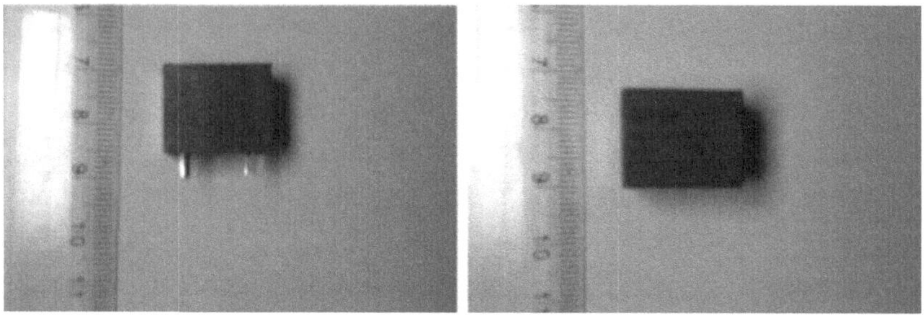

Fig. 2

Now, we see one most common circuit to control a relay, *i.e.,* using a transistor. Refer Figure 3.

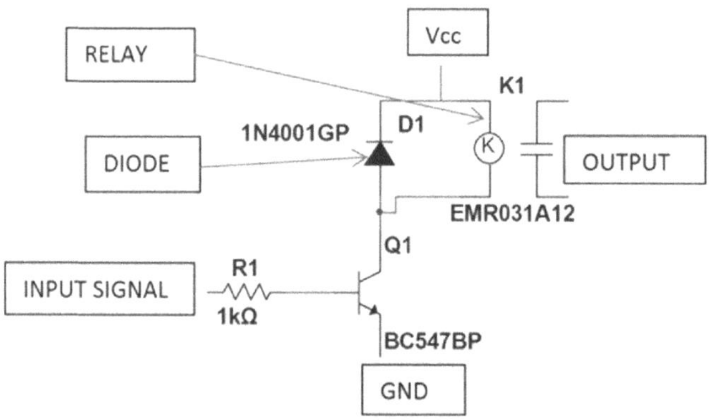

Fig. 3

The base signal controls the switching of the transistor. Once the base-emitter voltage is 0.7 volt, the transistor is in ON state and it conducts. Thus, current flows through the relay coils, making the armature move from NC to NO position.

You must have observed that a diode is placed at the two terminals of the coil. This is placed to prevent the back emf of the coil (which is essentially an inductor) from damaging the other circuit components.

Timer Circuits—IC555 as Monostable and Astable Multivibrator

IC555 timer is one of the most used integrated circuits in many electronic circuits. Its unique design and functionality has numerous applications. Here we are going to discuss its Monostable and Astable modes of operation. Figure 1 shows the 8 pin packaging of LM555 timer.

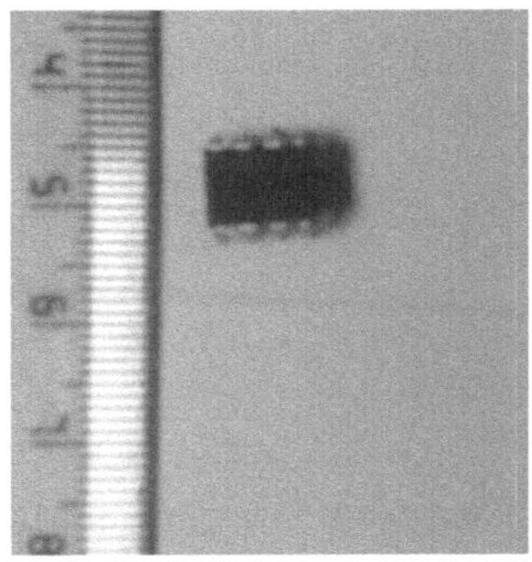

Fig. 1

Monostable Operation

Figure 2 shows the use of the IC in monostable mode. You can refer to the pin configuration of the IC from the datasheet supplied by the manufacturer.

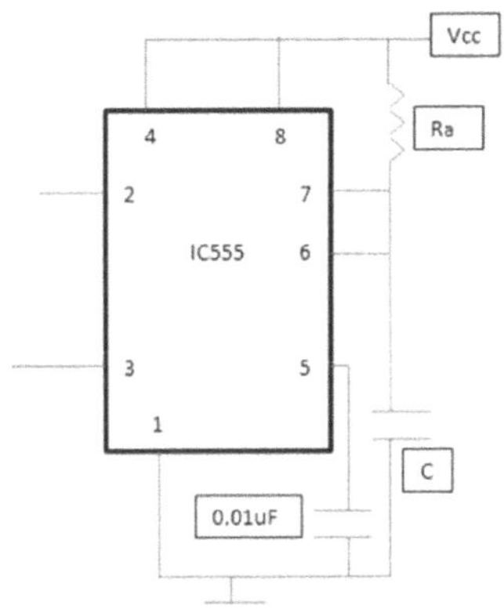

Fig. 2

In this mode, one needs to know about the trigger pin 2, resistance R_a and the capacitance C. When the pin 2 is grounded, or is kept at a potential below 1/3rd V_{cc}, the IC gets triggered. It produces a pulse of width T = 1.1 (Ra .C). Hence, the name is given as monostable. The output is obtained at pin number 3.

Astable Operation

Figure 3 shows the use of IC555 as an astable multivibrator. The circuit produces a chain of pulses with fixed ON and OFF time. These times are decided by the resistors Ra, Rb, and by the capacitor C.

Ton = 0.693 (Ra + Rb).C

Toff = 0.693 (Rb).C

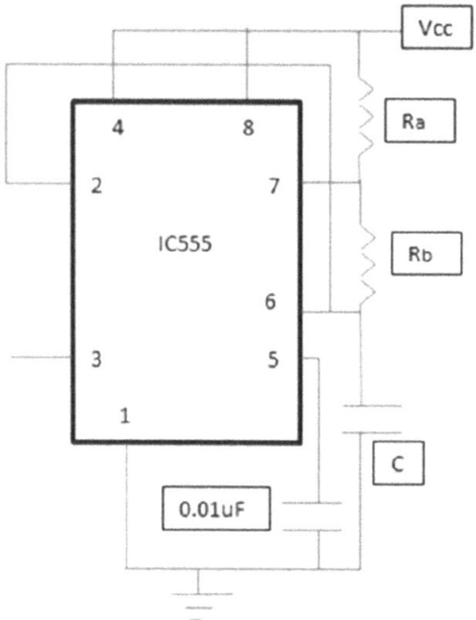

Fig. 3

Thus, we can get a square pulse of required high time and low time by choosing appropriate values of resistances and capacitances.

We get the output at pin number 3. We can calculate the other parameters such as total time period, frequency, duty cycle, etc. from the ON and OFF times calculated using the aforementioned formulae.

Waveforms

Monostable Mode

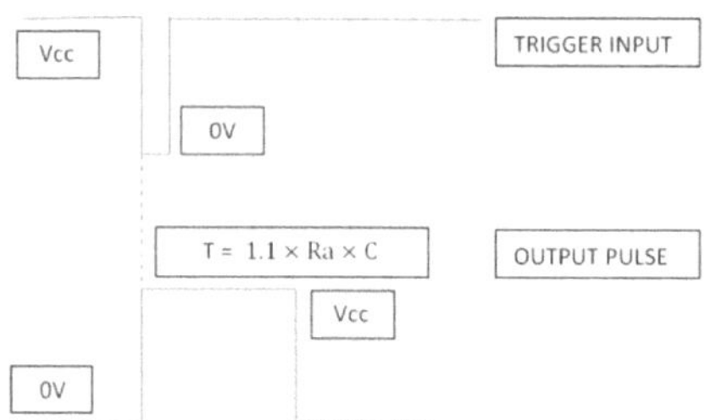

Astable Mode

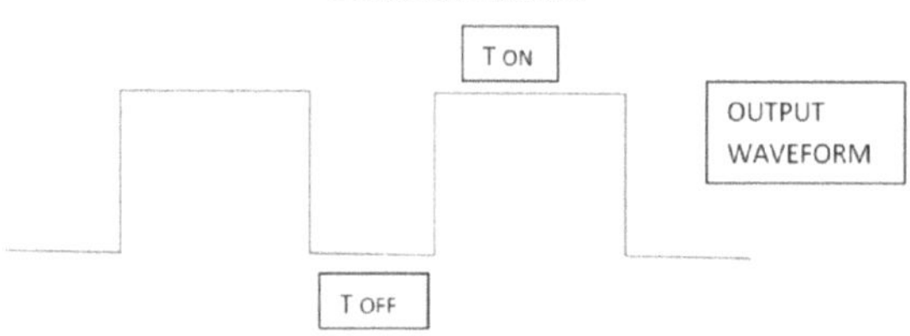

OP-AMPS

OP-AMPS or Operational amplifiers arc another commonly used set of ICs other than IC555. These have numerous applications and are used to develop complex integrated circuits. They are capable of performing arithmetic operations, both linear and non-linear, and hence have become widespread.

We first move on to understand the basics of an operational amplifier. Figure 1 shows the basic symbol of an operational amplifier.

There are various pins shown in the figure. Pin number 2 is the inverting input, pin number 3 is the non-inverting input, pin number 6 is the output, pin number 7 is the VCC and pin number 4 is V_{EE}.

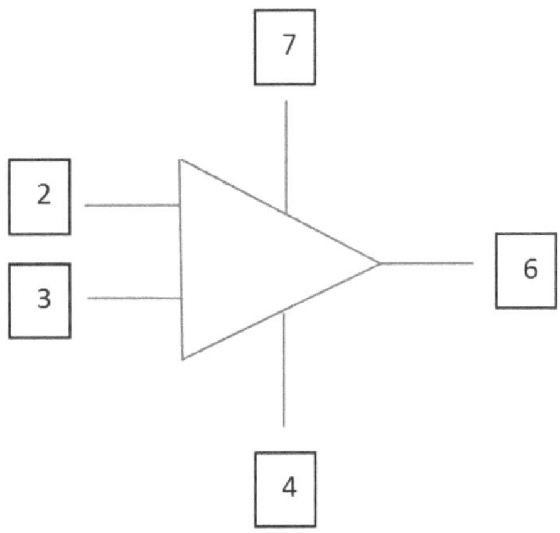

Fig. 1

There is one another term involved with OPAMPS known as gain. This decides the factor by which the OPAMP will amplify the difference between the two inputs. The output is given by

$$VO = A((V+) - (V-))$$

There are various characteristics of an OPAMP. And each of its characteristics has its specific application. We now proceed to see its pin diagram (Figure 2).

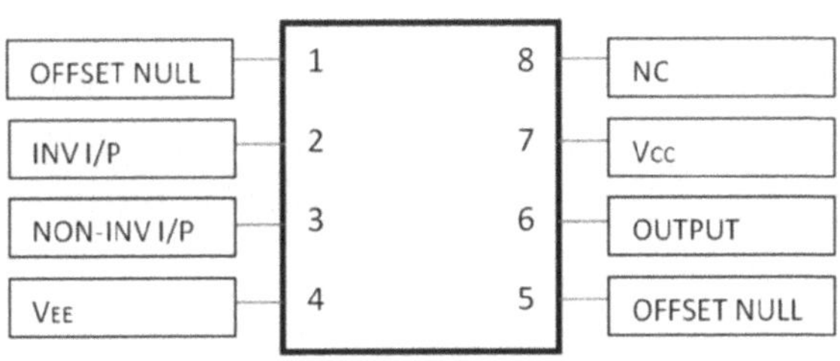

Fig. 2

The VCC and VEE connections are usually given +15 and –15 volts respectively. The non-inverting input is applied to terminal 3 and the inverting input is applied to terminal 2. The pin 8 is not connected and pin 6 is the output. The pins 1 and 5 are termed as offset null setting pins. A potentiometer is connected across these two terminals and it is adjusted to remove an unwanted factor called DC offset voltage.

Properties of an Ideal Opamp

1. Open loop voltage gain is infinite
2. The input impedance is infinite and the output impedance is zero
3. It has zero contribution to noise
4. The DC offset voltage is zero
5. It has infinite bandwidth
6. The voltages at both the input terminals remain constant. This gives rise to the concept of virtual ground, wherein if one of the terminals is grounded, the other has 0 potential.

Another terminology related to OPAMPS is saturation voltage. Though we apply source voltage of ±15 volts, we do not get 15 volts but a value slightly lesser than it (when the OPAMP is used in open loop, or the output voltage of closed loop is greater than 15 volts).

This voltage is the saturation voltage and defines the limit up to which we can amplify the voltage of the input. In the amplifiers to come, you must remember this fact.

OPAMP as an Inverting Amplifier

The term inverting amplifier basically means that the OPAMP is getting the input at the inverting terminal, the non-inverting terminal is grounded and a negative feedback is chosen. The device altogether is acting as an amplifier.

Since, we know that the OPAMP has ideally infinite input impedance, it does not intake any current. Secondly, since the non-inverting terminal is grounded, there is a virtual ground established at the inverting terminal.

We can thus write the two equations:

$$\frac{V_{in} - V(-)}{R_{in}} = \frac{V(-) - V_{out}}{R_f}$$

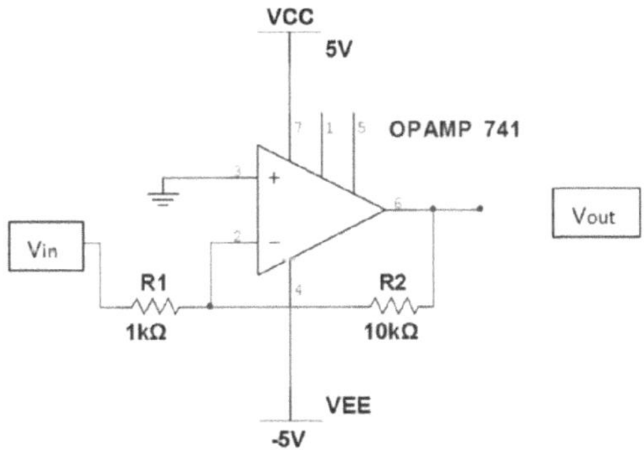

Fig. 3

As V (–) = 0

We get,

$$\frac{V_{out}}{V_{in}} = -\frac{R_f}{R_{in}}$$

Note: If both the resistances are same, then there is no amplification in the output, just a phase reversal. This application is termed buffer, and is generally used when impedance matching is required (When output current of the device is too low to drive further network, we use impedance matching circuits).

OPAMP as a Non-Inverting Amplifier

Figure 4 shows the OPAMP used as a non-inverting amplifier. Essentially, its inverting input is grounded, a negative feedback is used, and the input is given at the non-inverting terminal.

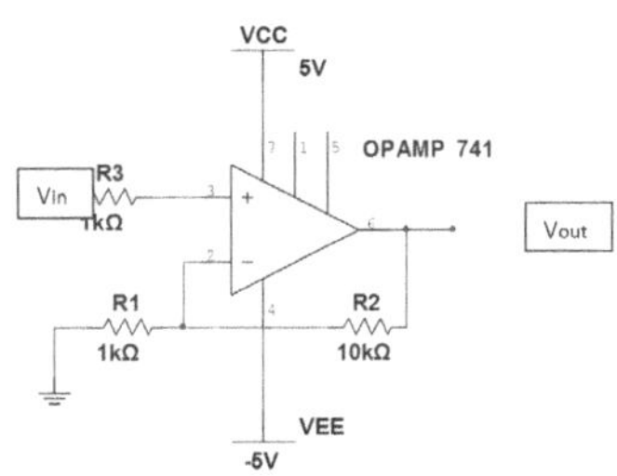

Fig. 4

The gain equation is obtained by following a similar pattern. The difference between the two voltages gives us the output. So we know one of them is Vin. The other one is $V_{out} \times \frac{R_f}{R_{in}}$.

We thus get the output voltage as:

$$V_{out} = V_{in}\left(1 + \frac{R_f}{R_{in}}\right)$$

OPAMP as Comparator

OPAMP basically is comparing the two inputs. When we use it in open loop, the gain is ideally infinite.

Therefore, the output is either $+V_{sat}$ or $-V_{sat}$, depending on the inputs.

Assuming both the potentials to be positive, we can use this device to set a reference point for the comparator, so that it gives a high output when the comparator exceeds the preset value (even the other way is possible).

This device is commonly used with transistors.

A potential divider based network is used to provide the reference potential at the inverting terminal, and the output of the sensor is applied at the non-inverting terminal.

As soon as the output voltage of the sensor exceeds that of preset, the output of the OPAMP goes to $+V_{sat}$.

Note: For TTL based logic levels, the two voltage levels, $+V_{sat}$ and $-V_{sat}$ are not suitable. So we either provide V_{cc} = 5 volts and V_{ee} = 0 volts or ground. We can also use other OPAMPS. LM358 is another commonly used OPAMP.

The following circuit shows the aforementioned use of comparators.

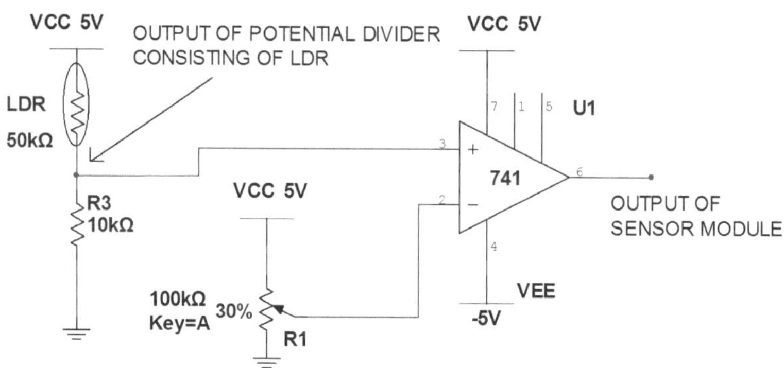

Fig. 5

This circuit is a light based triggering system, wherein we can set a particular voltage level, such that above a certain light intensity, the output of circuit is high.

This circuit is used in designing the light follower robot covered later in the book.

Thus, we have understood some basic applications of OPAMPS.

There are numerous other applications of OPAMPS; some of them include Differential Amplifier, Instrumentation Amplifier, Adder, Subtractor, Integrator, Differentiator, Current to Voltage converters, Voltage to Current converters, etc. You can try finding new applications or build circuits using the existing ones.

Optoisolators

These are devices that, as the name suggests, isolate one part of electric circuit from the other. These are generally used when certain devices have fixed range of acceptable inputs and get easily damaged by excessive currents.

Optoisolators are basically LED-photodiode pairs. If the LED goes high, the output of photodiode also goes high.

This principle is used to connect and disconnect one part of the circuit to another. And in case there is excessive current in the input side, there won't be any damage to the output side, and vice-versa. 4N35 and ULN2803A are some of the most commonly used optoisolators.

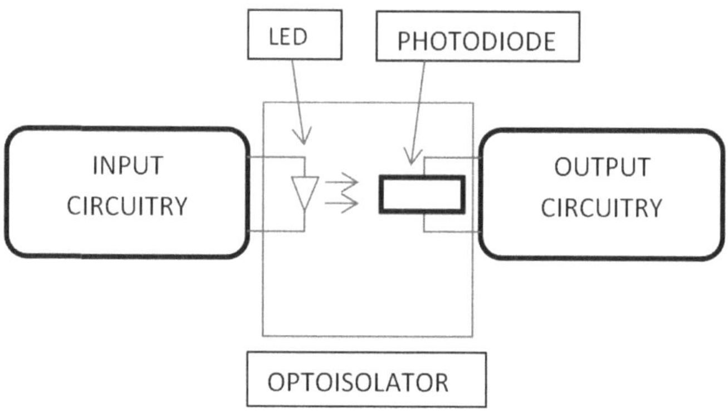

Fig. 1

Oscillators

Oscillators are devices that produce a chain of pulses without any trigger. The most commonly known oscillator is the RC Oscillator. The RC oscillator has no input but produces output pulses continuously, but it needs power supply.

Today, in various controllers and processors, Quartz crystal oscillators are used. These are high speed oscillators and do not require any external power. They only require two specific capacitances for proper operation.

The figure below shows a standard oscillator that is used in most of the circuits.

Fig. 1

Sensors

There are various devices/sensors that are used today in industries. These sensors measure various quantities and provide information about the particular quantity in real time in either analog or digital form.

These devices are usually interfaced with signal conditioning circuits or microcontrollers, and a control action is established on a particular task taking the sensor output as a feedback.

We now discuss a few of the sensors most commonly used in various projects and designs in industry today.

LDR (Light Dependent Resistor)

LDR as it is abbreviated is a variable resistor whose resistance is controlled by surrounding light. They usually have a light resistance of 1 KΩ and a dark resistance of about 1 MΩ.

They are made of either cadmium selenium or cadmium sulphide. LDRs have high resistance when there is no light falling on them, and vice-versa.

They are used to detect light having intensity comparable to that of an LED. They cannot be used to detect pulses of light of high frequency as they have a high response time.

Hence, they are generally used in applications involving low frequency of source.

Figure 1 shows a commonly used LDR.

Fig. 1

We now understand one of the most basic applications of LDR—switching of transistor by light. See the circuit (Figure 2).

Here, assume that there is no light falling on the LDR, hence, the resistance of the LDR is high (say around 2 M ohms). Hence, we can safely say that the base is grounded, base-emitter voltage is less than 0.7 volts, and hence the transistor is in OFF state.

If light falls, the potential of base rises, base-emitter voltage exceeds or is equal to 0.7 volt, and thus the transistor conducts (Figure 2).

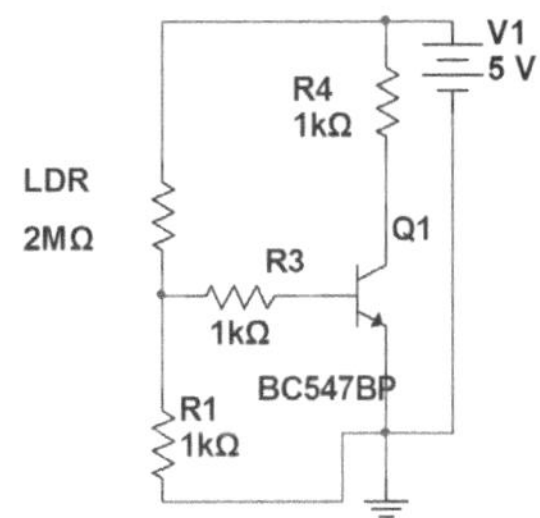

Fig. 2

Note: How can you obtain the operation in the opposite manner, *i.e.* the transistor should conduct when there is no light? You can achieve this by interchanging the positions of LDR and resistor R1.

Thermistor

A thermistor is a sensor that is used to measure temperature. It gives a resistance output and from the output we can determine the temperature of the medium it is immersed in.

Thermistors have negative temperature co-efficient, *i.e.* the resistance drops with rise in temperature. Thermistors are available in various packaging and temperature ranges, and we choose a particular one depending on the application it is required.

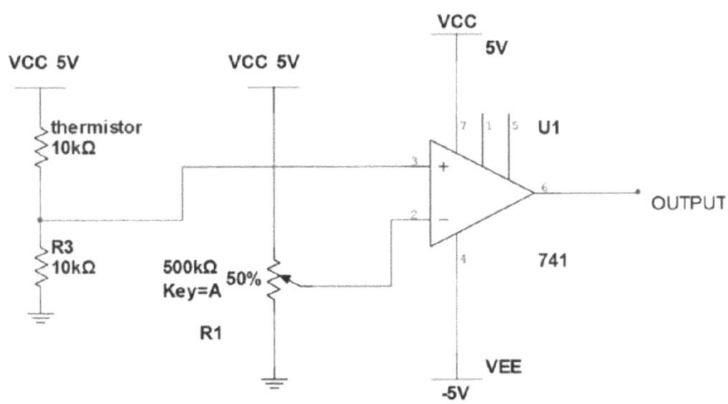

Fig. 3

Figure 3 shows a basic OPAMP based temperature alarm. The thermistor and R3 form a potential divider network. The output from lower end of thermistor is given to the positive input of LM741. At the negative end, a potential divider is used to adjust the set point. The output will go to positive saturation if V+ > V–. We can keep a buzzer or any further control circuitry at the output.

LM35 Based Temperature Sensor

Usually in electronic circuits, the bulky temperature sensors such as thermistors are not preferred to be used. Instead we use LM35. It is a semiconductor temperature sensor which operates with a +5 volt DC on 0 V reference. It is a 3 pin temperature sensor device in which a 5 V voltage source is given to the first pin, the second pin is grounded and third pin is taken for $V_{OUT.}$

This sensor gives a rise of 10 mV for every degree rise in temperature, and it has an output of 250 mV at 25 degree Celsius. It is generally preferred when the medium is non-hazardous in nature. These also have a good temperature range.

The LM35 series of temperature sensors are rated to operate over a –55°C to 150°F temperature range. These sensors do not require any external calibration and the output voltage is proportional to the temperature.

Working/USE of LM35

LM35 is an analog sensor that converts the surrounding temperature to a proportional analog voltage. The output from the sensor is connected to an analog to digital convertor to derive the equivalent temperature value in digital format.

Figure 4 shows the most common packaging of LM35 used in circuits. There are various other packaging available with more range of operating and storage temperatures.

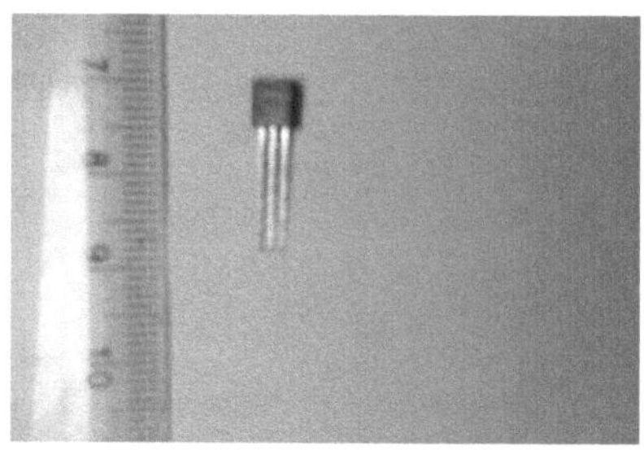

Fig. 4

IR – Photodiode Based Proximity Sensor

The most commonly used design of a proximity sensor available in market is shown in Figure 5.

Fig. 5

As you can see, it has a potential divider network, a comparator IC (LM385), IR LED, photodiode and two resistors. Its range is about 10 to 15 cms. It has three pins, one for supply (5 volts), and another for ground, and the last one for output. Based on the preset resistance on POT, the sensor gives a high output for the particular distance proximity. These sensors are commonly used in the field to robotics, to design line follower, obstacle avoider, wall follower, cliff avoider, etc. robots.

Working of the Proximity Sensor

An IR LED is a type of light emitting diode which emits infrared radiations. A photo resistor is a type of photo dependent which has a high output resistance in absence of light. In this circuit LM358 is being used as a comparator between the output of the photo resistor (to the non-inverting terminal) and a reference voltage set by a potentiometer (to the inverting terminal).

The output of the comparator is then connected to a diode.

Now a $+V_{CC}$ is obtained when the IRLED and photo resistor face a white platform and zero when they face a black platform. The Figure 6 shows a basic circuit diagram for the following module.

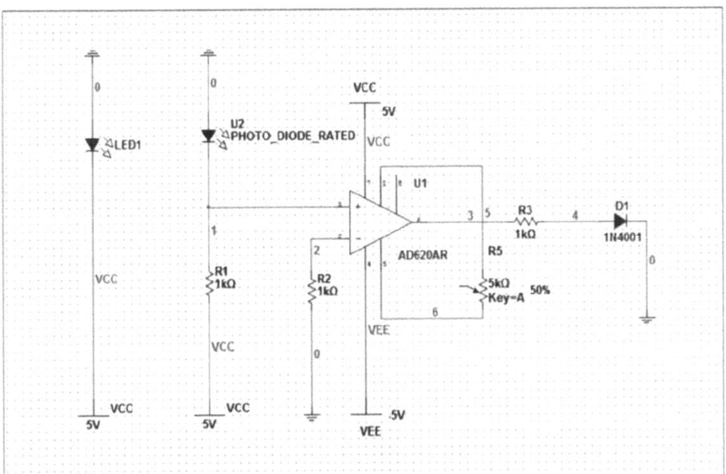

Fig. 6

Ultrasonic Eye

This is the most common distance measurement tool used today in the field of robotics. This sensor module has various components and five pins.

The two cylinder shaped structures are ultrasonic wave transmitter and receiver. The five pins are—Vcc, ground, trigger, echo and output.

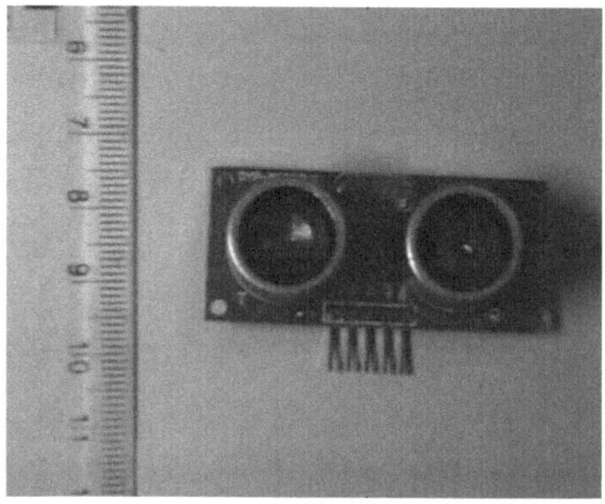

Fig. 7

In regular operation to find the distance in cms, we do not use the output pin. We provide 5 volts supply and connect the ground pin to ground. Next, to start the operation, the trigger input must be provided with the 38 KHz impulse input. Once applied, the sensor sends out an ultrasonic wave. Meanwhile, the echo pin gets set high as soon as the wave is sent.

When the receiver receives the wave, the pin turns low. We measure the transition time from high to low of the trigger pin to find the distance between the sensor module and the obstacle or wall. Its range varies from 250 to 300 cms.

A dedicated microcontroller is used to carry out the aforementioned operation. This module is commonly used in advanced robotics such as in hybrid or advanced obstacle avoidance robots.

MIC/Sound Sensor

A *microphone* is an acoustic-to-electric transducer or sensor that converts sound into an electrical signal. The electric signal from the microphone is fed into a JK or SR Flip flop circuit working in toggle mode. The output

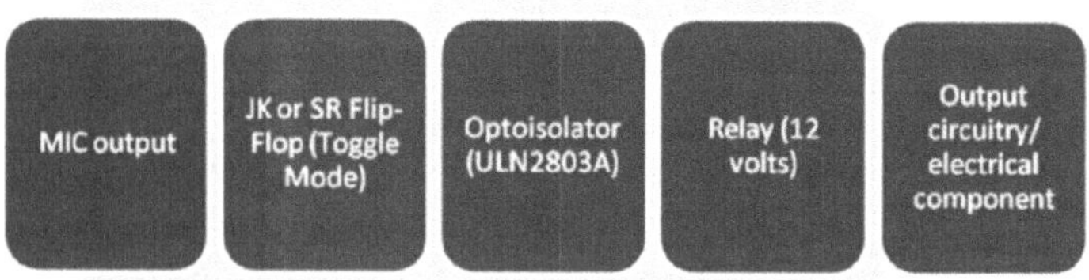

Fig. 8

from the flip flop circuit is fed to a 12 V relay. Since the output voltage of Flip flop is small for the relay coil voltage, the output from the flip flop circuit is fed Optoisolator 4N35 or ULN2803A which enhances the output voltage of the flip flop to a desired voltage level comparable to relay coil range. This relay output in desired configuration is connected to an electrical component for required applications. The above figure shows the flow of signal and the various stages in the sensing process starting from the MIC.

Hall Effect Sensor

These sensors are three pin devices, and look similar to transistors in their packaging. The figure shows a commonly used Hall Effect sensor. These sensors are used to detect presence of magnetic fields in the vicinity.

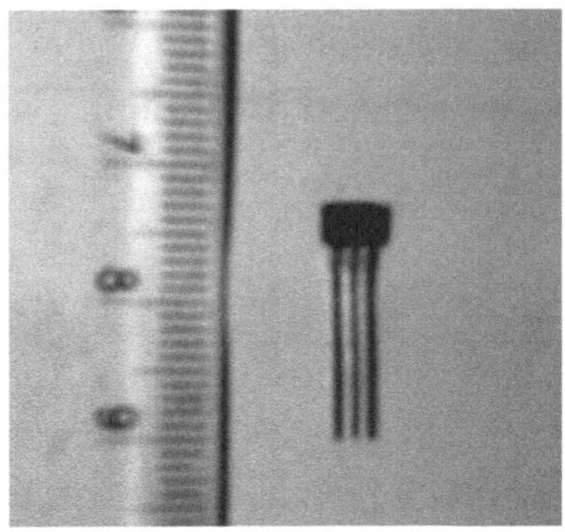

Fig. 9

The three pins of the sensor are—Vcc, ground and output. They are mainly classified into two types—active high and active low respectively. In the active low configuration, the output of these sensors is high in absence of magnetic fields and in presence, the output goes high. The concept is the opposite in active low configurations. HI300 is one commonly used Hall Effect sensor. Since these sensors are usually operated in TTL logic, a pull up resistor is used at the output to enhance the efficiency of the output.

Applications of this device are numerous, but one main application is the measurement of RPM/ speed in automobiles.

Gas Sensors

The main ones available in the market include MQ 1 to 7 series. These sensors are mainly based on the compound Tin Oxide. These sensors usually operate at elevated temperatures, and hence, each of these devices have inbuilt, low resistance heater coil, which heats up the main sensing element. Because of such a design, these sensors usually need about 500 mA of current.

The main principle of these devices is the change in output resistance. This property is tapped using a comparator design; a preset value of voltage is set, and the output of this sensor is used to control the voltage output of a potential divider arrangement.

A major disadvantage of these sensors is cross sensing, *i.e.*, a sensor is sensitive to more than one kind of gases and gives a high output for all of them.

Humidity Sensor

These sensors measure humidity by virtue of change in resistance or change in capacitance. If it is resistance based, then the same comparator based circuit is used, whereas if it is capacitance based, then IC555 timer is used (Its output waveform analyzed to determine change in capacitance).

Pressure Sensor

These sensors measure the differential pressure between the two pressure inputs. The pressure exerted induces a change in resistance. Strain gauges usually form the internal elements of these devices.

Vs+ and Vs– differential pressure output is obtained. Internal resistance of these devices is generally 100 ohms. Differential Amplifier gain system is used to make signal conditioning more accurate and efficient.

The output of these devices is usually in mV (around 10mV). One commonly used pressure sensor is NX10DP.

Flex

A flex sensor physically looks like a flexible plastic ruler that can bend easily without breaking. The bending movement changes the resistance at the output terminals.

It is usually used in safety systems of buildings.

Accelerometer

ADXL335 is one of the commonly used 3 axis accelerometer (x, y, z axis). These are mainly used as tilt sensors and provide an analog output.

Combinational Logic Design

Combinational logic is based on Boolean algebra. This kind of design is used generally to find the relation between the input and output quantities, and correspondingly build the circuit using logic gates.

It should be noted that combinational logic design can only be used in systems which are not time dependent, which don't have a memory and which don't have feedback!

To proceed with combinational logic design, you need to be clear with the concepts of logic gates dealt in the earlier sections.

If you are not clear, please revisit the earlier pages, read through once, and then proceed with this section.

There are few steps one needs to follow in developing the combinational design of a system. You need to understand as to how the relation between the inputs and the outputs is found out. We first write the relation between the inputs and the outputs in the form of a table.

Consider the following set of inputs A and B, and the outputs W, X, Y, Z.

Inputs		*Outputs*			
A	*B*	*W*	*X*	*Y*	*Z*
0	0	1	0	1	0
0	1	0	1	1	0
1	0	1	0	0	1
1	1	0	1	0	1

Note that 1 represents the high state and 0 represents the low state w.r.t. Transistor-Transistor Logic (TTL). We have the relation table and now we need the equations of the output quantities in terms of the inputs.

These equations can be of two forms—SOP (Sum of Products) and POS (Product of Sum).

Both these kinds of equations are essentially the same equations but are different in representation only. Let us understand how to obtain these equations from the table.

Understanding the SOP Form

This method essentially involves making the following table:

A	*B*	*Expression for Products*
0	0	$\bar{A} \times \bar{B}$
0	1	$\bar{A} \times B$
1	0	$A \times \bar{B}$
1	1	$A \times B$

We have obtained the product expressions by making all inputs high. If they are not high, we negate the input to make it high. Next, we need to consider the outputs one by one. Consider the output W; we know that W is high for (A = 0, B = 0) and (A = 1, B = 0). Thus, we write the corresponding products from the above table and add them. The equation so obtained is the SOP equation which gives us the relation between the inputs A, B and the output W. Hence we have,

$$W = \bar{A}\times\bar{B} + A\times\bar{B}$$

In a similar fashion, we can obtain the rest of the equations, *i.e.*, the expressions for X, Y and Z. They are as follows:

$$X = \bar{A}\times B + A\times B$$

$$Y = \bar{A}\times\bar{B} + \bar{A}\times B$$

$$Z = A\times\bar{B} + A\times B$$

Understanding the POS Form

This method essentially involves making the following table:

A	*B*	*Expression for Sums*
0	0	$A + B$
0	1	$A + \bar{B}$
1	0	$\bar{A} + B$
1	1	$\bar{A} + \bar{B}$

We have obtained the sum expressions by making all inputs low. If they are not low, we negate the input to make it low. Next, we need to consider the outputs one by one. Consider the output W; we know that W is low for (A = 0, B = 1) and (A = 1, B = 1). Thus, we write the corresponding sums from the above table and multiply them. The equation so obtained is the POS equation which gives us the relation between the inputs A, B and the output W. Hence we have,

$$W = (A + \bar{B}) \times (\bar{A} + \bar{B})$$

In a similar fashion, we can obtain the rest of the equations, *i.e.*, the expressions for X, Y and Z. They are as follows:

$$X = (A + B) \times (\bar{A} + B)$$
$$Y = (\bar{A} + B) \times (\bar{A} + \bar{B})$$
$$Z = (A + B) \times (A + \bar{B})$$

Simplifying the SOP and POS Forms

In this section, we not only will learn how to simplify the SOP and POS equations but will also understand that both equations are the same equations, *i.e.,* both are equal. But first, we need to recollect Boolean Algebra Laws. The various laws and rules are stated below:

(Note that the necessary ones are stated below. You can view the rest from various reference books).

The basic rules are as follows:

$X = 0$ or $X = 1$

$0.0 = 0$

$1 + 1 = 1$

$0 + 0 = 0$

$1.1 = 1$

$1.0 = 0.1 = 0$

$1 + 0 = 0 + 1 = 1$

Boolean algebra also has many laws. These laws have a property called Duality (Hence, the two equations in each law), wherein we replace AND by OR and vice versa, and 1s by 0s, and vice versa.

Commutative Law

(a) $A + B = B + A$

(b) $A.B = B.A$

Associate Law

(a) $(A + B) + C = A + (B + C)$

(b) $(A \times B) \times C = A \times (B \times C)$

Distributive Law

(a) $A \times (B + C) = A.B + A \times C$

(b) $A + (B \times C) = (A + B) \times (A + C)$

Identity Law

(a) $A + A = A$

(b) $A \times A = A$

Redundancy Law

(a) $A + A \times B = A$

(b) $A \times (A + B) = A$

(a) $A \times B + A \times \bar{B} = A$

(b) $(A + B) \times (A + \bar{B}) = A$

(a) $A + 1 = 1$

(b) $A \times 1 = A$

(a) $A + 0 = A$

(b) $A \times 0 = 0$

(a) $A + \bar{A} = 1$

(b) $A \times \bar{A} = 0$

(a) $A + \bar{A} \times B = A + B$

(b) $A \times (\bar{A} + B) = A \times B$

De Morgan's Laws

(a) $(\overline{A + B}) = \bar{A} \times B$

(b) $(\overline{A \times B}) = \bar{A} + \bar{B}$

As we are now aware of the Boolean Laws quite well, we can use them to reduce the SOP and POS equations:

We first deal with the SOP equations. They are as follows:

$$W = \bar{A} \times \bar{B} + A \times \bar{B}$$

$$X = \bar{A} \times B + A \times B$$

$$Y = \bar{A} \times \bar{B} + \bar{A} \times B$$

$$Z = A \times \bar{B} + A \times B$$

Consider the first equation:

$$W = \bar{A} \times \bar{B} + A \times \bar{B} = \bar{B} \times (A + \bar{A}) = \bar{B}$$

Similarly,

$$X = B \times (A + \bar{A}) = B$$

$$Y = \bar{A} \times (B + \bar{B}) = A$$

$$Z = A \times (B + \bar{B}) = A$$

We now deal with the POS equations. They are as follows:

$$W = (A + \bar{B}) \times (\bar{A} + \bar{B}) = (A + \bar{A}) \times \bar{B} = \bar{B}$$

In a similar fashion, we can obtain the rest of the equations, *i.e.*, the expressions for X, Y and Z. They are as follows:

$$X = (A + B) \times (\bar{A} + B) = (A + \bar{A}) \times B = B$$

$$Y = (\bar{A} + B) \times (\bar{A} + \bar{B}) = (B + \bar{B}) \times \bar{A} = \bar{A}$$

$$Z = (A + B) \times (A + \bar{B}) = (B + \bar{B}) \times A = A$$

We thus have finished simplifying the SOP and POS equations, and we can conclude that both these forms are same. On simplification, both give us the same equations.

You can apply both methods and try finding these equations for other input-output systems.

You now can obtain the relation between the inputs and outputs for any defined system. This technique will help you design these systems with logic gates, thus making the system cheap and efficient. It can help you define logical actions for various devices including robots such as Line Follower, Obstacle Avoider, etc.

Multimeter Basics

The Figure 1 shows the most basic and commonly used multimeter. The physical observation shows that it has two probes—one display and one selector knob.

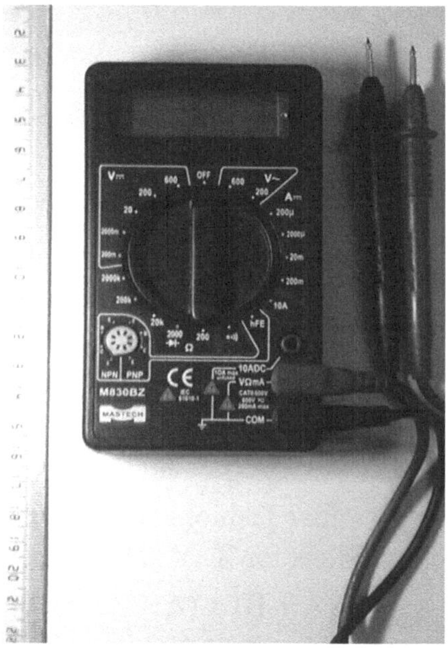

Fig. 1

It is capable of measuring a lot of electrical quantities such as resistance, voltage (DC and AC), continuity test, diode test, transistor test, current measurement, etc. the advanced models available today can even measure a certain range of capacitance, temperature, etc. Once you have clearly understood the basic view, we proceed to understand the proper use of selector switch and the probe terminals. Refer the figure on next page.

The V---sign shows DC voltage. The device is capable of measuring dc voltages of up to 600 volts. To measure 2 volts, we keep the selector on the nearest higher voltage level, so in this case it is 20 volts. Similar fashion, we adjust the selector for other quantities. The V symbol indicates measurement of AC voltage. It can measure AC voltages of up to 600 volts. Similarly, it can measure resistances from a few ohms to two mega ohms. The continuity test is to check whether two points in the circuit are electrically connected or not. This mode is useful especially when we are debugging circuits. For measuring voltage or resistance, the probes are connected to the terminals V and COM. Then we keep the pins in parallel to the device and obtain the measurement of the required quantity on the display.

Incase current needs to be measures, the red probe must be inserted into the 10 ADC terminal, and the probe pins must be connected in series with the path wherein the current needs to be measured.

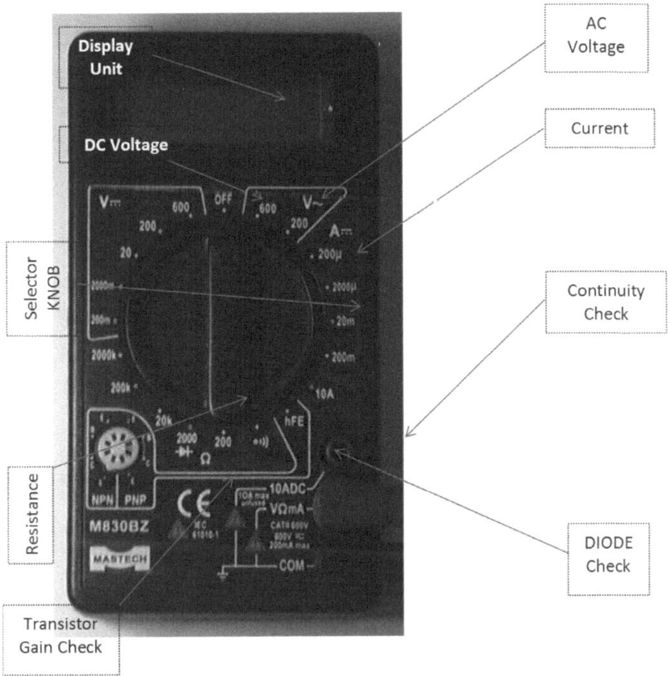

Fig. 2

Piezo-Electric Buzzer

A piezoelectric buzzer is one of the most common devices used in various circuits. It is a device which produces a loud sound when DC voltage of the correct polarity is supplied. Generally it is used to indicate an alarm when a particular control condition is satisfied. Some applications include water fill indicator, alarm clock, etc.

The objective is to provide a current of around 50 to 70 mA, which the control signal cannot provide. This much current is required for the buzzer to give off a loud sound. Hence, we use a transistor based driver circuit. The buzzer is driven by the collector current, and the control signal is applied to the base of the transistor. This is a simple example of transistor based switching.

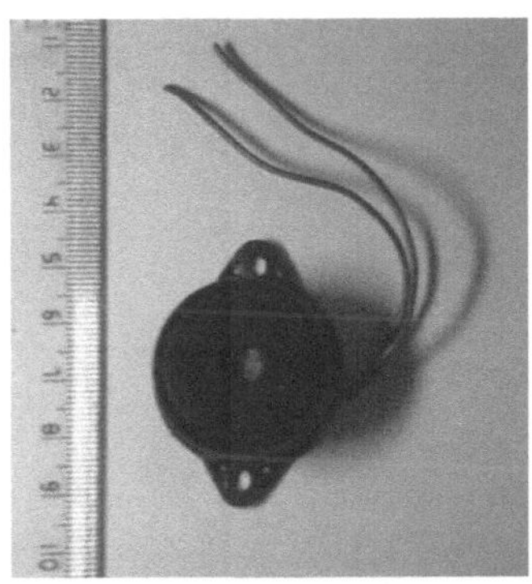

Fig. 3

Part II

Robotics

Manual Robot

Understanding of basic electronics is the basis of understanding robotics. We now move on to understand as to what is a manual robot.

What is a Manual Robot?

To be honest, it's a simple toy car whose motion can be controlled.

Why do you call a toy car a Robot, we were expecting the one's we see on TV…

A toy car fits into the definition of a Robot (which is any device with more than a degree of freedom) and thus is called one. It is the least an engineer could do.

What exactly should a toy car, (oops!) I mean Manual robot have?

We all know what a toy car needs:

- Motors + Wheels for movement
- Framework for support
- Power supply to power the Motors
- A device through which you could control the motion.

So, how do we go about all this?

It's said the best way to finish a number of tasks is to take them one at a time.

Subsystems of a Robot

Locomotion System

The locomotion system of the robot gives it translational and rotational degrees of freedom *i.e.,* it enables the robot to move. Elementary locomotion systems usually consist of motors, wheels and castors. There are different types of motors which can be used on any robot, the most common ones include:

- DC Geared Motors
- DC Motors without Gears
- DC Servo Motors
- DC Stepper Motors.

Power Supply System

The power supply (Figure 1) rating that the robot demands must be met by the power supply put on the robot. For that calculations are done as follows:

For series connections, find the sum of the voltage rating of the devices such as motors, etc. and for all parallel connections, the rated current is the sum total of all individual rated currents.

Fig. 1

Generally 12 volts, 500 mA is enough for most applications however for wireless 9V Li+ ION batteries low current rating can be used.

Control System

In this system (Figure 2), mainly the direction control of the DC motor is discussed. Most common way to control the direction of rotation is using a DPDT switch. The figure on the next page explains the same:

If C-E and D-F are connected the motor rotates in one direction. When E-G and F-h are connected, the polarity of the supply to the motor is reversed and hence, the direction of rotation is reversed.

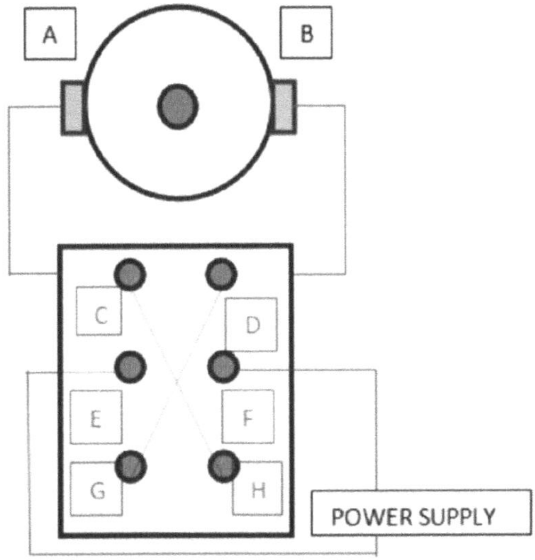

Fig. 2

Alternative Direction Control—Using L293D

In today's electronic circuits (Figure 3), instead of DPDT switches, dedicated ICs such as L293D are used. This IC can handle the direction control of two motors.

The L293D has four input pins and 4 output pins, passing appropriate sequence to the input pins result in desired motion of motor. The figure below shows the packaging of the IC.

Fig. 3

We now proceed to understand the pin configuration of the L293d and how various connections are made. Refer the figure on the next page.

The various pins of the L293D (Figure 4) are shown in the figure. The four inputs are the signals from the controller or the circuitry/keypad used to send the control signal. The output terminals are used to connect the DC motor to the IC.

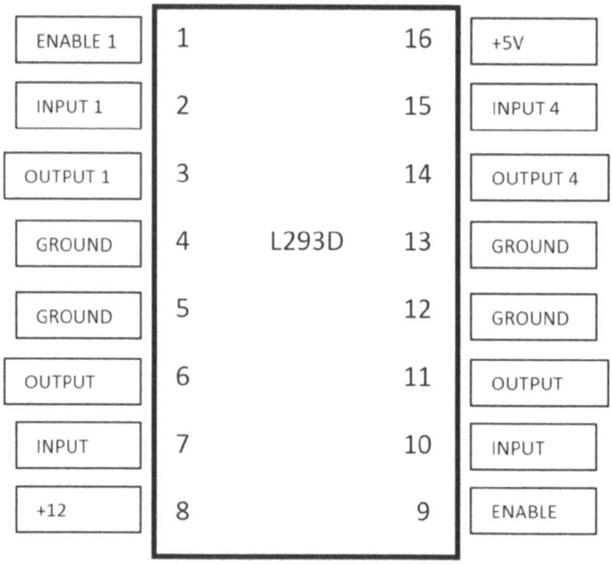

Fig. 4

The pins 8 and 16 are used to connect the two power supplies, one for TTL logic of the IC and the other to run the motors.

Enable pins are used to inhibit the motion of motors. If the enable pin is set low, the corresponding side motor gets turned OFF.

Fig. 5

This IC is common to almost any road robot because of its simple functionality and use. The Figure 5 shows a commonly available module of this IC.

Now, let W be first input, X be the second input, Y be the third input and Z be the fourth input. This convention is used throughout the manual.

We need to understand now how the motion is affected by these bits.

W	*X*	*Y*	*Z*	*Motion of The Robot*
1	0	1	0	Forward
0	1	0	1	Reverse
1	0	0	1	Right
0	1	1	0	Left
1	1	1	1	No motion
0	0	0	0	No motion

Refer the direction control of DC motor. We can now see as to how the various combinational set of inputs to the L293d module can control the way the robot moves. Remember this convention for all the future designs of robots you make.

Understanding H-Bridges

L293D is a modified or developed form of H Bridge. In general, the H-bridge is used for polarity interchange (Figure 6).

An H-bridge is nothing but a set of four switches, properly aligned. H-bridges can be realized using relays or transistors. The diagram below shows the implementation using transistors. A similar approach can be taken to develop it using relays.

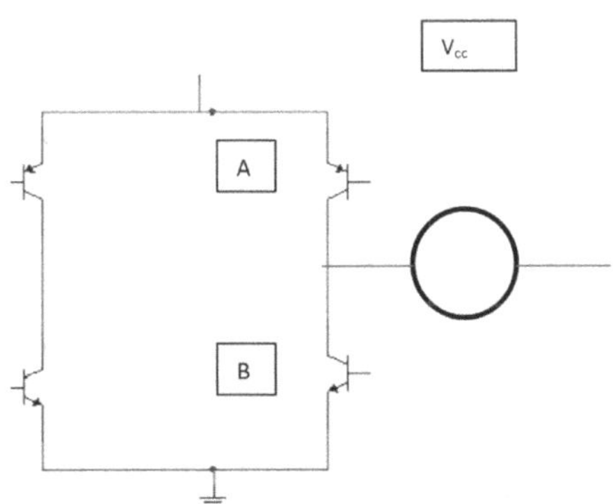

Fig. 6

Consider A, B, C and D as the four inputs as shown in the figure.

A	*B*	*C*	*D*	*Function*
1	0	0	1	Forward
0	1	1	0	Reverse
1	1	0	0	Brake
0	0	1	1	Brake
1	0	1	0	Fuse Test
0	1	0	1	Fuse Test

Here we have one NPN and another PNP transistor. The need for the change in transistor configuration is required since one of the end terminals of the transistors must be connected to V_{cc} and the other to the ground.

In NPN the emitter is usually connected to the ground and in PNP the emitter is connected to the V_{cc}. The above circuit essentially uses switching of transistors.

Block Diagram of a Manual Robot

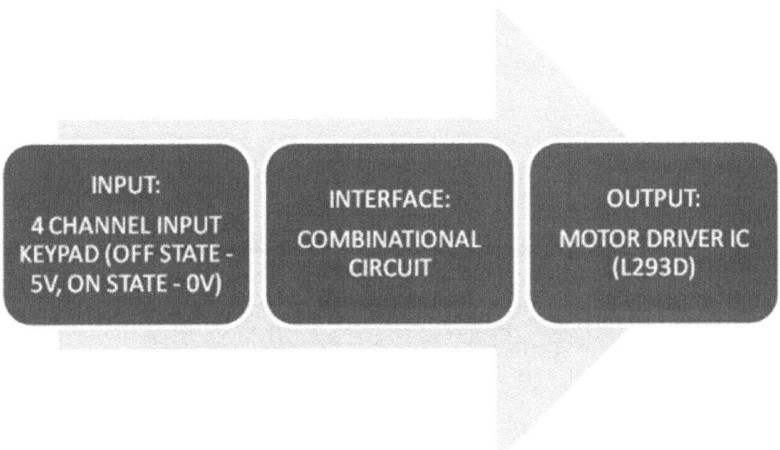

The above flow/process diagram shows the various stages in the manual robot. The first is the input stage wherein we use a 4 channel keypad to provide the required control signals.

Next stage is the combinational circuit that acts as the interface circuit to convert these signals into the required ones for controlling the motor.

The last stage is the motor driver unit that receives the signals and accordingly controls the direction of the rotation of the motors.

4 Channel Input Keyboard

The Figure 1 shows one such module/keyboard. Here the switches are connected in active low configuration, such that if a switch is pressed, the corresponding output will turn low.

The switches are numbered 1, 2, 3 and 4 as shown in the adjoining figure.

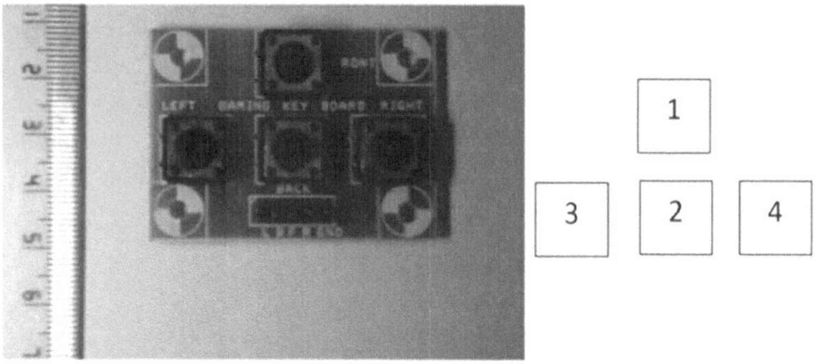

Fig. 1

Now we require the following output conditions for the inputs (as shown in Table 1).

Table 1

1	2	3	4	O/P
0	1	1	1	FRONT
1	0	1	1	BACK
1	1	0	1	RIGHT
1	1	1	0	LEFT

Note: Since entire combinational logic design is based on TTL, we can use a device known as an LED array to test the output of a system. This device has a certain number of LEDs (say 8), and hence, can test eight lines at a time. It needs a ground connection. The Figure 2 shows an LED array.

Fig. 2

Interface Circuit

Since L293D needs inputs like (1010), (0101), (1001), (0110) we need an interface circuit. Let 1, 2, 3, 4 be inputs and A, B, C, D be outputs. Recollect the combinational logic design procedure explained earlier in the book. The same is to be followed here in Table 1.

Table 1

1	2	3	4	A	B	C	D
0	1	1	1	1	0	1	0
1	0	1	1	0	1	0	1
1	1	0	1	1	0	0	1
1	1	1	0	0	1	1	0

For all the other input combinations, the output needs to be 0000. Using the SOP form we get:

$$A = 1'.2.3.4 + 1.2.3'.4 = (1 \oplus 3).(2.4)$$

$$B = 1.2'.3.4 + 1.2.3.4' = (2 \oplus 4).(1.3)$$

$$C = 1'.2.3.4 + 1.2.3'.4 = (2 \oplus 3).(2.3)$$

$$D = 1'.2.3.4 + 1.2.3.4' = (1 \oplus 4).(2.3)$$

So we require 1 XOR IC and 2 AND ICs for the combinational design Figures 1 and 2.

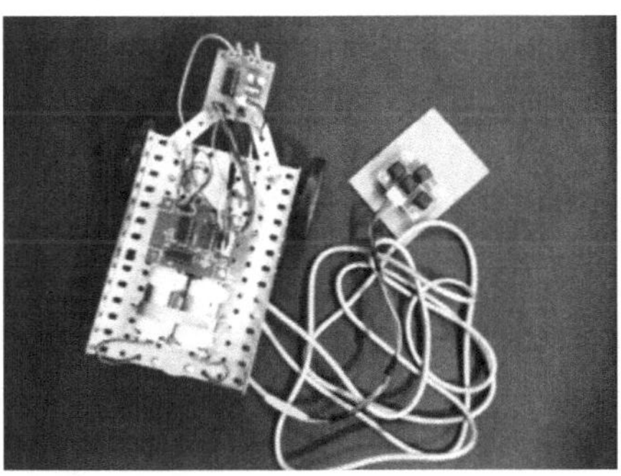

Fig. 1

Fig. 2

RF Based Wireless Manual Robot

What is a Wireless Manual Robot?

So as you have guessed there is no direct point to point contact between the sources of data from the user to the destination, which happens to be the robot itself in our case.

The transmitter (source) and the receiver (destination) must have separate power supply; otherwise you wouldn't have the right to call it a wireless robot.

Well ideally in an open environment one can get ranges up to 20–30 meters and by integrating what we call a satellite antennae the directionality of the signal can be enhanced and the range can be increased up to two fold.

What exactly should a wireless Manual robot have?

1. Everything a manual wired robot has except the wires from it.
2. A mechanism to transmit the output data bits from the four channel keyboard. The mechanism is realised using:
   - Encoder (HT12E)
   - RF Transmitter 433 Mhz
   - Decoder
   - RF Receiver 433 Mhz.

So Let's have a look at the block diagram (Figure 1) depicting the entire process for a better understanding.

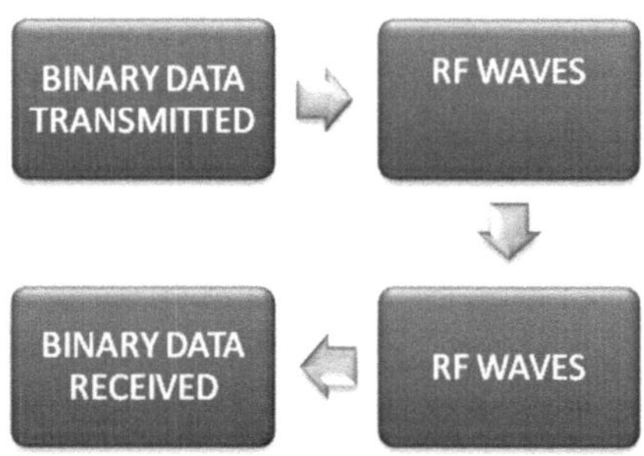

Fig. 1

Block Diagram of RF Based System

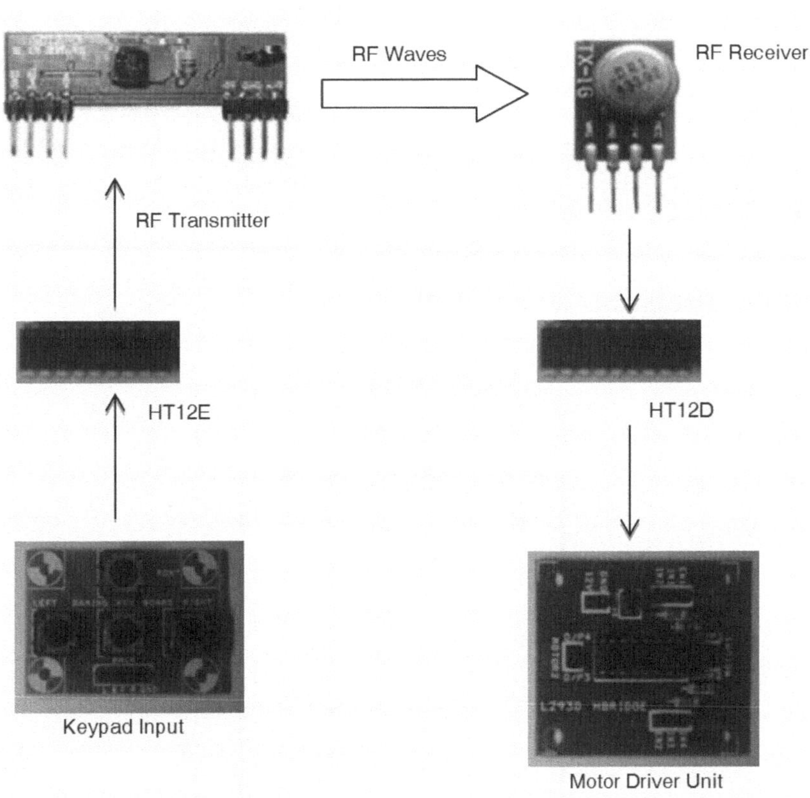

Encoder and Decoder

Encoding means converting from one form of signal or data to another.

Consider a situation where a French and a German need to communicate. Since neither of them understand each other's language, they would require an interpreter so that each of them can understand each other's message. Similarly many ICs function on different protocols and need this interpreter as an interface. The ENCODER does the interpreter's job.

The decoder also converts one form to another, but usually it converts the signal/data to the original form as given to the encoder. It can be said as the interpreter who does the reverse process of encoder. The encoder is used at the transmitting end whereas the decoder is used at the receiving end.

The encoder and decoder HT12E and HT12D are both 18 pin ICs and are available in DIP packaging also in the market. The next two pages show the connections that are to be made between the encoder and transmitter, and decoder and receiver.

The AD 11, 10, 9, 8 are the data bits at the transmitting side, wherein we connect the keypad. The D 11, 10, 9, 8 are the data bits which are output from the decoder.

The Transmission Side Circuitry

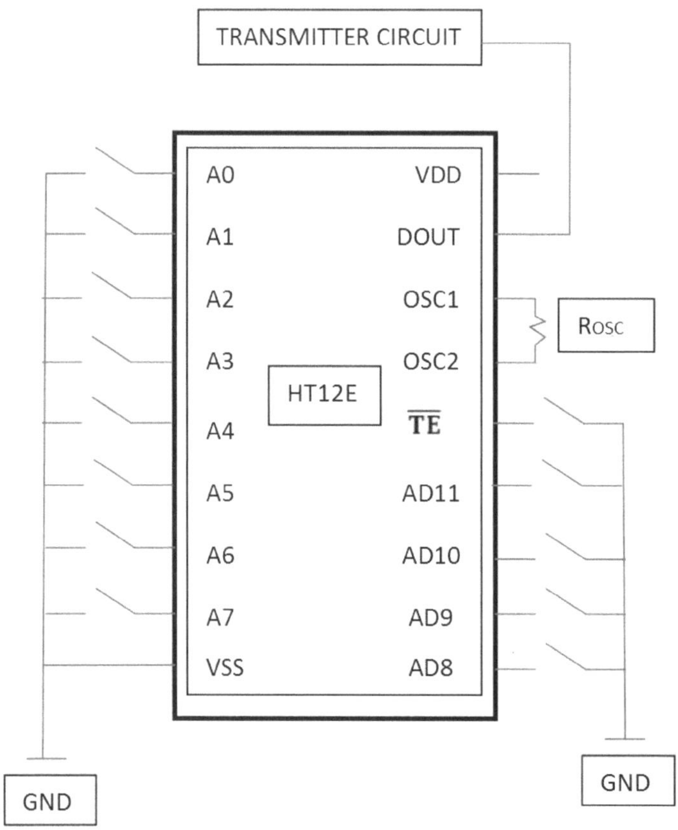

The Receiver Side Circuitry

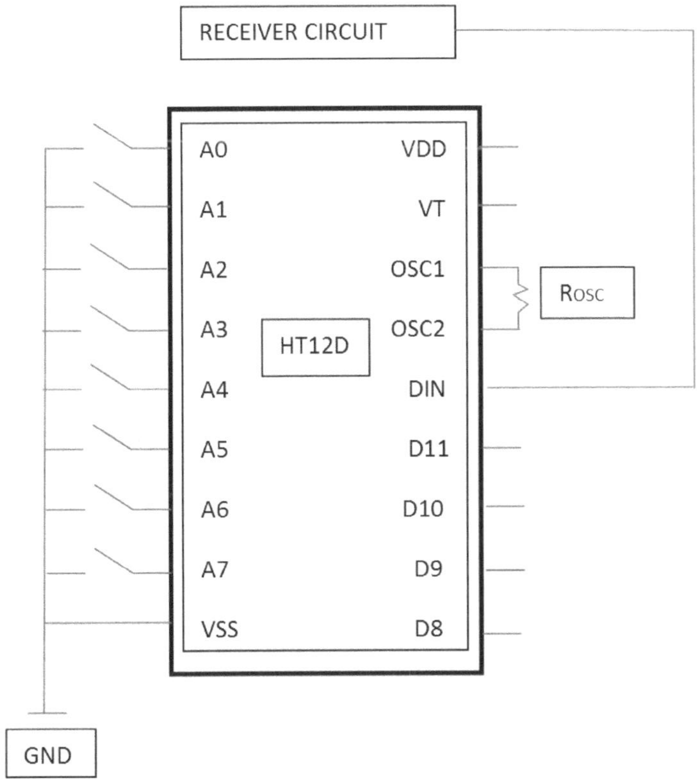

RF Transmitter and Receiver

The transmitter and receiver operate at 433 MHz. These devices send/ receive information serially at an extremely fast BAUD rate (bits transmitted per second).

The data transfer is PISO (parallel input serial output) at the input and then SIPO (serial in parallel out) at the output. There is a provision slot for connecting a satellite type of antennae.

The Figure 1 shows the commonly used RF transmitter and receiver modules.

Fig. 1

Working of the Circuit

The HT12E-12D series of encoders and decoders transmit 12 bits of data out of which 8 are address bits and 4 are data bits. For both the encoder and the decoder A0–A7 represents address bits and D0–D4 represents data bits.

For the decoder circuit, the Oscillator resistor to be connected is 51 KΩ and for the Encoder circuit, the oscillator resistor is 1.1 MΩ.

Only when the Transmit Enable pin of the encoder is grounded the encoder sends data to the transmitter. The encoder is Active Low, which means it sends a high signal when not under operation *i.e.,* by default all pins of HT12D will be high.

HT12E has 8 address pins for giving a fixed address for the communication. This address makes the communication having a unique code to transmit data. The receivers having the same unique address code can receive the data bits.

The data bits are taken parallel into the four input data pins. A combinational series data bits are generated at the Data output pin. This occurs@433 Mhz and hence we are not able to see any data lack.

The data is then transmitted to the RF Transmitter and is received by The RF receiver which feeds the Decoder through the Data Input Pin serially. This received data is then converted into parallel data bits which are forwarded to a combinational circuit and can be used for required application.

Enabling Security in the RF System

What is the need to enable security in a wireless system?

The transmission of RF waves happens at a particular frequency from an allowable set of frequencies (bandwidth) depending on the values of crystal oscillator and air core inductor/capacitor used in the circuit of the transmitter. So what if someone else too transmits at the same frequency? How would your receiver recognize that someone was trying to jam/prevent the data transmission? And presume the "someone else" also transmits 12 data bits, how would your receiving circuit choose which to accept?

So, clearly there is a need of some sort of security in this network.

So how exactly do we enable security?

Luckily for HT12E-12D series of encoder and decoder the answer lies in the working mechanism itself. As we know that a 12 bit data is transmitted so the first eight bits which are the address bits needs to match on both the transmitting side and the receiving side. So these address bits act like a sort of password, until they match, no data is accepted.

The safest and a very high level implementation would be to assign a common clock to both the transmitter and receiver circuits and keep changing their address bits in a predetermined sequence or even use a sequence generator.

Mobile Controlled (DTMF Based) Wireless Robot

So what is a Mobile Controlled Robot?

As the name suggests the robot can be operated via the keypad of the mobile phone. Which means sitting at home I can control a robot placed in antarctica provided the service provider has a good network in both regions. So we see the range is no longer limited.

Sounds Exciting!

Let us first have a brief understanding of the working mechanism

The Keypad of the mobile phone transmits the pressed key on the keypad via a form of signal known as DTMF. The key pressed on the keypad is generally heard as a sound on the receiver's end so if we did connect a compatible audio jack we would have the audio signal as output from the audio jack.

Now our task is to process this signal and convert this audio signal into some form of binary signal such that the output be different binary combinations for different keys pressed on the keypad.

This task is done using what we call an encoder and luckily the output is the binary form of the number pressed itself. CM8870 is the encoder used for this purpose.

The Figure 1 shows how the CM8870 can be used to convert DTMF tones into binary values.

So once the output is obtained from the receiver mobile phone then we can design a combinational circuit and use it for the desired applications.

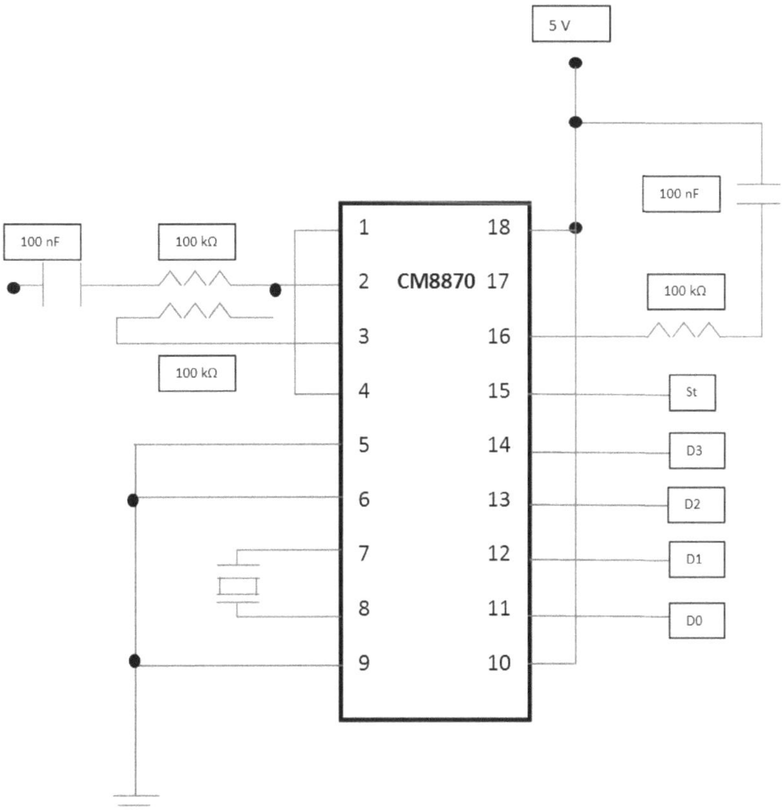
5 V
100 nF
100 nF
100 kΩ
100 kΩ
100 kΩ
100 kΩ
CM8870
1
2
3
4
5
6
7
8
9
18
17
16
15
14
13
12
11
10
St
D3
D2
D1
D0

Fig. 1

Block Diagram

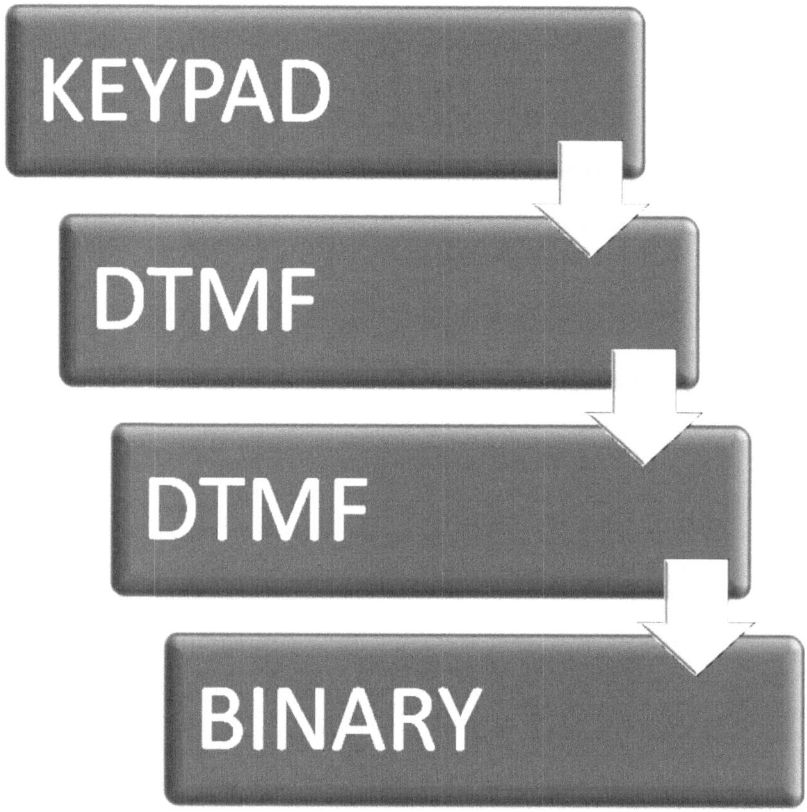

DTMF and Working of the System

DTMF signal is used over analog telephone lines in the voice frequency band between telephone handsets and other communication devices and the switching center.

DTMF is a method of instructing switching system of the telephone numbers to be dialed, or to issue commands to switching systems.

Robotic control using the DTMF is achieved when the user at the transmitting side pressed mobile phone keypad buttons and at the remote location receiver receive the signal so that robots takes corresponding actions.

DTMF Decoder IC- HT9170B

- 18 pin IC.
- Operational voltage- 2.5 V–5.5 V.
- One wire of the headphone is connected to the wire of IC and another to ground.
- Connect the cellphone (receiver) to 3.5 mm jack at other end.
- Keep the cellphone in auto answer mode.
- Now call on that phone from another phone (transmitter).
- Now press the keys on the remote phone and you will see your robot moving forward (press 2), backward (press 8), turn left (press 4), turn right (press 6).
- Four output pins (D0, D1, D2, D3) of the IC is connected to any port of the microcontroller.
- DTMF (dual tone multiple frequency) technology is very useful in home automation *i.e.,* with the help of this technology we can control our home appliances.

Fig. 1

Fig. 2

Fig. 3

Designing of Some Robots Using Combinational Logic Design

Designing an Obstacle Avoidance Robot

We will be designing an obstacle avoidance robot based on IR-photodiode proximity sensors. The circuit will consist of three main parts:

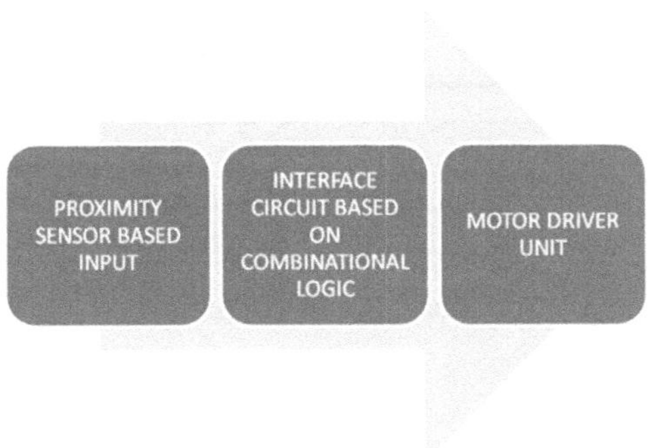

The output of the proximity sensors will be used to determine the direction of motion of the robot. This will be done by suitably converting the output into proper control signals for the motor driver unit.

This flow will remain the same for any of the robots discussed in the section. Only the sensor and the interface circuit change from design to design.

We first start with discussion of how the proximity sensors are used and placed on the robot chassis. Consider Figure 1.

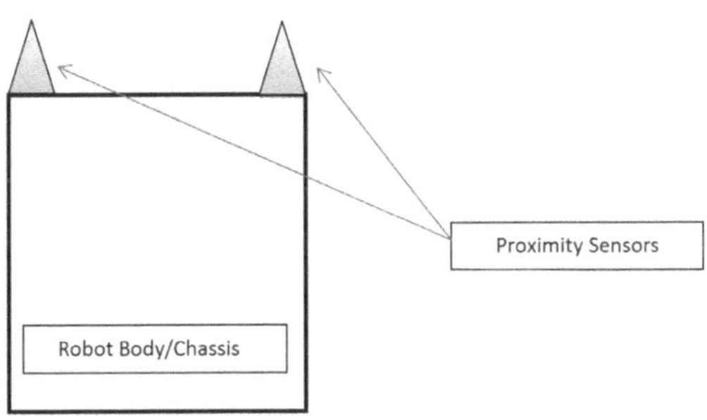

Fig. 1

Let the left proximity sensor's output be A and the other output be B. We follow the standard approach in designing the combinational logic interface.

In the following design, we are designing the robot such that if it encounters an obstacle, it turns right, else it goes straight. We are trying to implement the simplest possible design of this robot. We now draw the input-output in Table 1.

Table 1

A	*B*	*W*	*X*	*Y*	*Z*
0	0	1	0	1	0
0	1	1	0	0	1
1	0	1	0	0	1
1	1	1	0	0	1

W, X, Y and Z are the outputs from the combinational logic interface which act as the inputs of the motor driver unit. By the same approach of SOP or POS, we obtain the equations as:

$$W = 1$$

$$X = 0$$

$$Y = \overline{A + B}$$

$$Z = A + B$$

The following circuit can be easily implemented using logic gates. We will require only one NOT gate and one OR gate. Thus, the robot will work in the manner we desired.

It is to be noted that it would be better if the range of the proximity sensor is kept maximum so that the robot has sufficient distance to stop and turn.

Another thing to be noted is that the robot will not make a smooth turn. As soon as it encounters an obstacle, it will turn till the sensor outputs turn low. Then again it will slightly ahead, till the outputs turn high. Again the robot will turn. This cycle repeats but usually the robot is able to turn in about two such cycles.

Fig. 2

Fig. 3

Fig. 4

Designing a Line Follower Robot

The same proximity sensors are used for the line follower robot. But here, the application is slightly different.

In this case we make use of the fact that the sensor gives a high output when placed in front of a white object, and gives a low output when placed in front of a black object.

Hence, we design the interface circuit, and we use two sensors placed in line with each other, and facing the bottom.

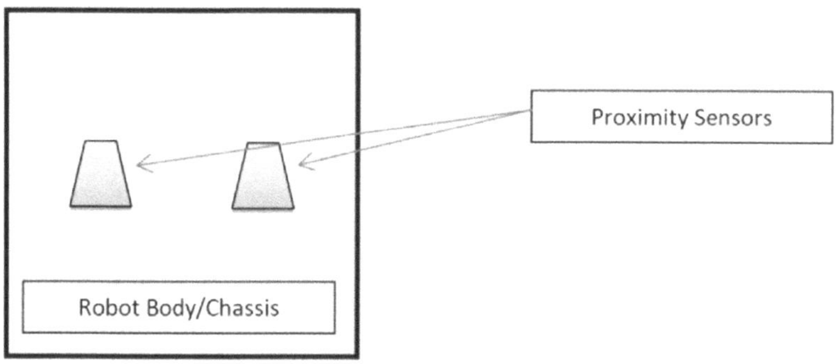

Fig. 1

We use a black line, thick enough to come in proximity of both sensors when in straight line. Refer the figure below. The sensors are placed at the bottom of the chassis.

When the black path is straight, both the sensors have low output. We want that in this case the robot keeps moving straight.

In next case, when there is a right turn, the left proximity sensor will go out of the black path at a certain point of time as the robot was initially moving straight. We want the robot to turn right in this case.

Similarly, for the left turn. And in the case when both the inputs go high, *i.e.,* the robot is out of path, we set it to come backwards as there are chances the robot may overshoot out of the track at some steep turns.

We thus compute the table of inputs and outputs—the first step for the combinational design.

A	*B*	*W*	*X*	*Y*	*Z*
0	0	1	0	1	0
0	1	0	1	1	0
1	0	1	0	0	1
1	1	0	1	0	1

By using the SOP or POS approach, we get the various outputs as follows:

$$W = \bar{B}$$

$$X = B$$

$$Y = \bar{A}$$

$$Z = A$$

We can thus develop the interface circuit using logic gates, and the outputs of this stage act as inputs to the motor driver stage. We will require only two NOT gates to design the following logic.

It is to be noted that the distance between the proximity sensors is critical in designing the robot. If we keep the sensors close (around 1 cm), the robot will travel faster on straight lines (both the sensors have low O/P), but it will travel slower and more zig-zag on turns.

If the distance between the sensors is kept more, the robot will turn faster but will take time on straight line paths.

Designing a Light Follower Robot

A light follower robot also follows the same structural pattern as seen in obstacle avoidance robot. The input is taken from LDR based sensor module. The output is then processed using the combinational logic design and then the output of combinational logic is fed to the motor driver stage.

The Figure 1 shows the basic structure of this robot. The LDR modules have the LDRs facing upwards.

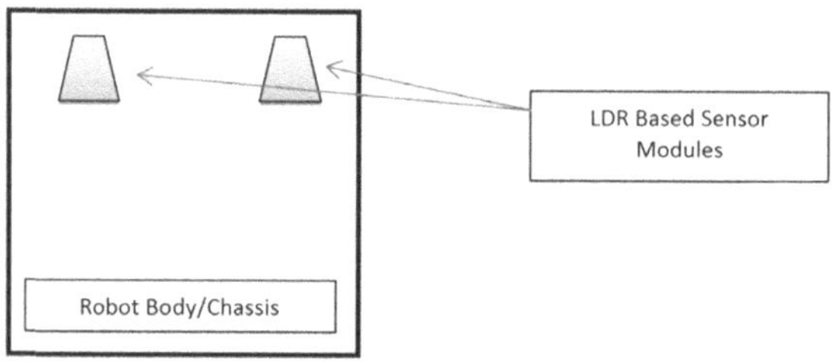

Fig. 1

We first discuss about the LDR based sensor module developed for the purpose. Refer Figure 2.

The LDR needs to be tested first and three observations need to be made initially. The three observations are:

- Rdark or resistance of the LDR in dark
- Rambient or resistance of LDR under ambient light conditions
- Rtorch or resistance of LDR when illuminated with a light source or torch.

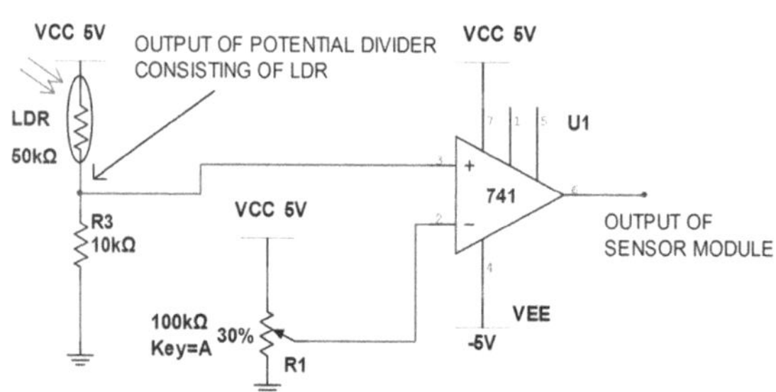

Fig. 2

It is to be noted that the ambient light conditions should not be broad daylight. The normal room lighting will be ideal for the ambient condition resistance measurement.

Secondly, note that while measuring resistance of LDR under illumination or ambient conditions, the light source is not close to the LDR. This is done for better functioning of the robot.

For the test LDR, the three values obtained are as follows:

Rdark = 145 MΩ

Rambient = 4.45 KΩ

Rtorch = 1.2 KΩ

Now, based on these values we choose an appropriate resistance to form the potential divider with LDR. We take a 10KΩ resistor for the purpose. Under torch conditions, the voltage output is 4.464 volts and, under ambient conditions, the voltage output is 3.440 volts.

We want our sensor module to give a high output only when the torch is shone on it. We thus take the average of voltage outputs under torch and ambient light, and keep it as the reference potential input to pin 2. The reference voltage value thus obtained is 3.95 volts.

We have designed the sensor module. We now need to design the required combinational logic interface.

Designing of the Combinational Logic

We have to design the combinational logic so that it satisfies the three conditions:

1. When both the outputs of sensors are low, the robot should be stationary.
2. When either side one sensor's output is high the robot should turn in the corresponding direction.
3. When both the outputs are high, the robot should move straight.

We hence prepare the corresponding table 1 showing the inputs and outputs of this stage.

We can solve for the outputs W, X, Y and Z using SOP or POS, and develop the circuit as a standard procedure. But, we will now understand as to how the design of line follower robot can be slightly modified to develop this robot's combinational interface.

Table 1

A	*B*	*W*	*X*	*Y*	*Z*
0	0	0	0	0	0
0	1	1	0	0	1
1	0	0	1	1	0
1	1	1	0	1	0

For that purpose, we again have a look on the combinational design of the line follower robot (Table 2).

A	*B*	*W*	*X*	*Y*	*Z*
0	0	1	0	1	0
0	1	0	1	1	0
1	0	1	0	0	1
1	1	0	1	0	1

Consider the case when A = 0 and B = 0. We want our robot to stop in this case, so we connect the outputs of both the sensors additionally to the enable pins to the motor driver IC L293D. One must recollect that if the enable pin of the motor driver is made low, the corresponding side motor will be turned off.

For the next three cases, we observe that if we interchange the outputs W and X, and Y and Z of the line follower design circuit (while connecting it to the motor driver module), we get the required combinational design for the light follower robot.

Make sure you make the connections with the motor driver unit correctly; else the robot will not function well.

Some additional criteria to be taken care of are the alignment of LDRs and the placement of sensor modules on the chassis of the robot.

The sensor modules should be placed such that when the torch beam is incident, it can cover both the LDRs, else the robot will never go straight.

Secondly, the angle which the LDR units make with the vertical axis also affects performance of the robot. If you incline the LDRs at 35 to 45 degrees with the vertical and inclined towards each other, the robot controlling will become easier.

Part III

Microcontrollers and Autonomous Robots

Microcontrollers and Autonomous Robots

What are Autonomous Robots?

As the name suggests it has something to do with autonomous and independence indeed. The independence happens to be with regard to the ability to process what it sees or feels and act as per these stimuli.

These robots are inherently dumb devices but when logic is embedded and when they are directed to as to what must be done in all possible events then they are at times more efficient than humans.

So one can comprehend that these robots vary from each other on the basis of their input sensing capabilities, processing speeds and response times.

But what is that critical element in a robot which gives it an artificial brain?

This device is the mother of all embedded systems and is called the microcontroller. As the name suggests the components inside must be of micro level and since it has the capability to control so controller. The block diagram of a microcontroller will help us understand its interior components.

Block Diagram of any Microcontroller

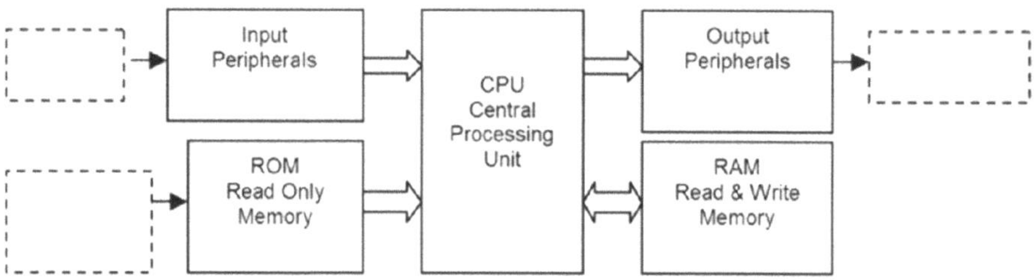

Now one needs to understand that it's just not enough to have a microcontroller; it requires some hardware and software development environment. This hardware development environment is called the development board. Freeware compilers like MPLAB's HI TECH-C from MICROCHIP itself serve as the software development environment.

Introduction to PIC Development Board

A well designed development board is always essential for smooth development/experimentation with microcontrollers. Every hobbyist must have some kind of developmental setup. Some people use breadboard while others use general PCB to fabricate their development system. Development is not very easy if bread boards or general PCB are used. If bread board is used, after few days of use, the wire will become loose or come out of the hole. While making large circuit in a general PCB is a tedious job.

In HPS's PIC Development board an attempt has been made to address the same. It is a circuit board with basic connection required to drive the microcontroller and also with features like serial port, prototyping area (LED, Switch), In-built power supply regulation and it supports 40 pin *PIC* microcontrollers.

A development board is very useful as you can program the microcontroller and also check the outputs using the LEDs or even connect it to the rest of the project using wires or connectors. The power supply provided on the board can be used to supply power to other circuits.

A very good advantage of such a development board is that all the basic project requirements are present on the board and it also occupies less space. So one can say the *PIC* development board is the perfect circuit to design, make and test very interesting and innovative projects.

Introduction to PIC Microcontroller Series

PIC is a family of Harvard architecture microcontrollers made by Microchip Technology, derived from the PIC1640 originally developed by General Instrument's Microelectronics Division.

The name PIC initially referred to "Peripheral Interface Controller". PICs are popular with both industrial developers and hobbyists alike due to their low cost, wide availability, large user base, extensive collection of application notes, availability of low cost or free development tools, and serial programming capability.

Microchip announced on February 2008 the shipment of its six billionth PIC processor.

The PIC architecture is characterized by its multiple attributes:

- Separate code and data spaces (Harvard architecture) for devices other than PIC32, which has a Von Neumann architecture.
- A small number of fixed length instructions.
- Most instructions are single cycle execution (2 clock cycles, or 4 clock cycles in 8bit models), with one delay cycle on branches and skips.
- One accumulator (W0), the use of which (as source operand) is implied (*i.e.,* is not encoded in the opcode).
- All RAM locations function as registers as both source and/or destination of math and other functions.
- A hardware stack for storing return addresses.
- A fairly small amount of addressable data space (typically 256 bytes), extended through banking.
- Data space mapped CPU, port, and peripheral registers.
- The program counter is also mapped into the data space and writable (this is used to implement indirect jumps).

There is no distinction between memory space and register space because the RAM serves the job of both memory and registers, and the RAM is usually just referred to as the register file or simply as the registers.

The architectural decisions are directed at the maximization of speed-to-cost ratio. The PIC architecture was among the first scalar CPU designs, and is still among the simplest and cheapest.

The Harvard architecture—in which instructions and data come from separate sources—simplifies timing and microcircuit design greatly, and this benefits clock speed, price, and power consumption.

Advantages of PIC Architecture

- Small instruction set to learn
- RISC architecture
- Built in oscillator with selectable speeds
- Easy entry level, in circuit programming plus in circuit debugging PIC Kit units available from Microchip.com for less than $50
- Inexpensive microcontrollers
- Wide range of interfaces including I2C, SPI, USB, USART, A/D, programmable Comparators, PWM, LIN, CAN, PSP, and Ethernet.

LED Interfacing

In this section you will be learning how to interface a LED to a microcontroller. This is one of the most basic tasks that one can be done using a microcontroller. By interfacing a LED properly one can learn the functions of a port in a microcontroller.

A port is a channel capable of receiving input or transmitting output.

To make a port function for the required application, knowledge of TRIS registers and PORT registers is required. By making user defined header files for simple projects like this, one can learn the basics of user defined header files.

Let's start with a simple task of switching on a LED.

A LED glows when the positive terminal of the LED (longer wire terminal) has a higher voltage with reference to the negative terminal (shorter terminal).

The difference in voltage between the two terminals should be at an optimum level of 3 V. Also to avoid Breakdown of LED a Current limiting resistor (220 Ω to 400 Ω) is always added in series with the LED.

In our development board this can be done in two ways, the longer terminal (positive terminal) can be connected to the 5 V pin and the shorter terminal can be connected to one of the pins of one of the ports and we may set that pin low OR, the shorter terminal (negative terminal) can be connected to the ground pin and the longer terminal can be connected to one of the pins of one of the ports and we may set that pin high.

Note: Do not forget to connect the series resistor!

To switch on a LED the first thing that should be done is, make the particular pin as output pin. Assigning value to the TRIS registers makes a port behave either as an input or an output port. Assigning a high value to TRIS register makes a port input and assigning a low value makes it an output port.

How do we do that?

It's simple!

PIC16F877A has 5 ports namely PORT A, B, C, D and E. Each port has 8 pins with a few exceptions A(6), E(3). All these ports can be made input or output. Now in this project we will look to interface the LED by using the first pin of PORT B, *i.e.,* RB0.

Also one may note that PORT B of PIC16F877A has the feature of internal pull up resistors which is controlled by a register called RBPU (active low). An internal pull up resistor removes the need to connect a series resistor with the LED.

So now the task is to make PORT B as output. This can be done in two ways:

1. *You write logic LOW to the whole port.*

This is done as follows:

```
TRISB = 0x00;
```

0x00 is in hexadecimal and in binary it is

0000 0000.

It means all 8 pins of PORT B are made as output pins. But a disadvantage is that all the 8 pins have been assigned to be output port and they can't be used for any other task (*i.e.,* to act as input pin).

So we follow another method.

2. *You write logic LOW only to the particular pin.*

This is done by masking the PORT & is logic AND operator. We know that anything AND operated with 1 will retain the previous value and anything AND operated with 0 will always be 0. So by performing AND operation on TRISB bit RB0, no matter what the previous value is, it becomes 0, thus making it an output pin.

Now that the port or pin is made as output, it is time to write logic HIGH into the port and make it give a logic HIGH *i.e.,* +5 V output to glow the LED.

```
TRISB &= 0xfe
```

This is also done in the same way as TRISB, but the only difference is that instead of TRISB we will be using PORT B.

```
PORTB = 0xff;          or          RB0 = 1;
```

Thus LED is turned ON by giving logic HIGH to PORT B pin RB0. Similarly to switch OFF LED, the PORT is still output port but logic LOW should be written into the port. So TRISB will be the same and you need not write it again. Logic LOW is written into Port B pin RB0.

```
PORTB = 0x00;          or          RB0 = 0;
```

By writing 0x00, the whole port *i.e.,* all 8 pins will go LOW. So the second method is used. So the LED is switched OFF by writing logic LOW into the Port.

Now that we learnt how to switch ON and switch OFF a LED, let's put it in a small program such that the LED switches ON for some time an switches OFF for some time.

For this we need to use delay function. By giving a certain amount of delay after switching ON the LED, the LED will remain ON for that much amount of time and then it can be switched OFF and with a delay it will remain OFF for than much amount of time.

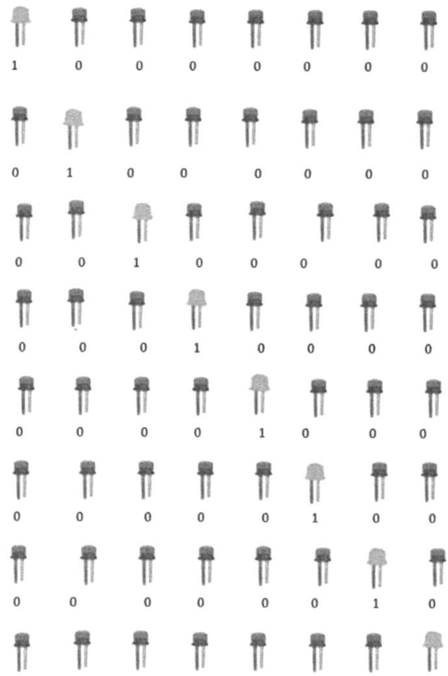

Main.c

```
#include "main.h"
    void main()
    {
    TRISB=0x00;             //Configures PORTB as an OUTPUT port
    PORTB=0x00;             //Initializes PORTB with LOW on all pins
while(1)
        {
                PORTB=0b10000000;    //makes RB7 high
    Delay(1000);
                PORTB=0b01000000;    //makes RB6 high
                Delay(1000);
                PORTB=0b00100000;    //makes RB5 high
                Delay(1000);
                PORTB=0b00010000;    //makes RB4 high
                Delay(1000);
                PORTB=0b00001000;    //makes RB3 high
                Delay(1000);
```

```
                PORTB=0b00000100;    //makes RB2 high
                Delay(1000);
                PORTB=0b00000010;    //makes RB1 high
                Delay(1000);
                PORTB=0b00000001;    //makes RB0 high
                Delay(1000);
        }
}
```

Main.h (The file to be included in every odd code)

```
#include <htc.h>//standard header file storing all commands
#include "Delay.h"//header file for delay function

//DEFINE MACROS
__CONFIG (0x2F0A);  //configuration bits set as per requirement

//FUNCTION PROTOTYPES
void main(void); //will change depending on program
```

Delay.c (The same code is used almost in all programs)

```
#include "delay.h"
void Delay(unsigned int time)
{
    unsigned char pause;
    while( time > 0)
    {
            pause = 255;
            while(pause--);
            time--;
    }
}
```

Note

The delay.c, delay.h and main.h files which are shown above are used throughout this manual. So if these delay.h is included in any project, then it means that the code written above is used.

Troubleshooting

Sometimes the LED may not glow and reasons for that may be:

- The LED may be connected in wrong direction
- Check the anode and cathode properly and connect the LED
- You may not be getting enough voltage
- Using multimeter check if you are getting at least around 2 V. If not then check if there is any break in the track and fix it
- The TRIS value may not be set properly
- Set the TRIS value to 0 and make sure that LED is connected to that pin only. Sometimes the LED may be connected to other pins.

Switch Interfacing

In this section you will be learning how to interface a switch and microcontroller. This is yet another very basic project that one can do using a microcontroller. This again helps in teaching the basics of microcontroller programming. Let's start with a simple task of making the Switch pin as input port.

Usually in all developments boards a LED is connected to the 34th pin *i.e.,* PORT B pin RB1. So now the task is to make PORT B (RB2) as input. This again can be done in two ways:

1. *You write logic LOW to the whole port.*

This is done as follows:

```
TRISB = 0xff;
```

0xff is in hexadecimal and in binary it is 1111 1111.

It means all 8 pins of PORT B are made as input pins. But a disadvantage is that all the 8 pins become input port and they can't be used for any other task (*i.e.,* to act as output pin). So as usual we follow the other method.

2. *You write logic LOW only to the particular pin.*

```
TRISB |= 0x04;
```

| is OR operator. We know that anything OR operated with 0 will retain the previous value and anything OR operated with 1 will always be 1. So by performing OR operation on PORT bit RB2, no matter what the previous value is, it becomes 1, thus only pin RB2 goes high.

Now that the port or pin is made as input, it is time to enable the internal pull up resistors. So the port gives 5 V and when the switch is pressed the pin gets grounded and thus we can check the condition of the switch accordingly.

This is also done by setting low to the RBPU register.

```
RBPU = 0;
```

So now how do we check if the switch is pressed or not? For that we use the conditional statement.

```
If(RB2==0);
```

This means that as the internal pull up is enabled, when switch is pressed the particular pin is grounded and it goes to logic LOW *i.e.,* it is cleared. So by checking if the bit is clear or not we can know if the switch is pressed or not. The conditional statement returns 1 if bit is clear and a zero if it is not. Now that we know the basics of switch and LED interfacing, we can write a program using both.

Switch.c

```
#include "main.h"

main(void)
{

    TRISB &= 0xFD;          //Led as output port
    TRISB |= 0x04;          //Switch as input port
  RBPU = 0;                 //Enable internal pull-up resistor

    for(;;)                 //Loop for ever
  {
            if(RB2 == 0)    //Check for switch status
                    RB1 = 1;    //If it is pressed, the led turns on
            else
                    RB1 = 0;        //Else led turned off
    }
}                                   //End of function
```

Switch Bouncing and Debouncing

When we press a switch, sometimes due to spring like nature of the contacts of the switch, the switch creates oscillations. If this is used in a digital circuit the oscillations will look like multiple switch presses and this can lead to errors in output. This is called as switch Bouncing. This is overcome by Switch Debouncing. It can be done mechanically or through software. In software it is done in the following way:

- First the condition is checked if switch is pressed or not. If it is not pressed the function will return 0.
- If the switch is pressed then it will give a very small delay of may be 1 ms and then again check if the switch is pressed or not.
- If pressed then return 1 or else it means that the initial detection was only oscillations.

Serial Communication

In this section you will be learning how to interface the PC to a microcontroller. This is will help you understand how data acquisition systems are made.

Communication with the PC is done via the Serial port of the microcontroller which is connected to the PC through the DB9 connectors to the USB jack via an interface and a cable.

The data from the controller can be analyzed by various software on the computer and be further processed using the same. Serial communication is also used to communicate with field devices. We thus start understanding from the basics.

There are mainly two types of communication modes—serial and parallel. The main difference between the two lies in the number of data lines used; in serial communication, only 1 data line is used.

The problem lies as to how the PC will know as to which of the bits are what? Basically there are loads of 0s and 1s received by the receiver and unless there is some common instruction set they share, it is impossible for the receiver to make out of the bits. We hence have something known as protocol, which defines the entire process of communication.

The process of serial communication is known as USART communication (Universal Synchronous Asynchronous Receiver-Transmitter). There are various terminologies one needs to understand first.

Start and Stop Bits

These bits are additionally transferred via the channel to mark the start and stop of serial communication data. The start bit initiates communication and in a sense tells the receiver to get ready to receive data.

The start bit is usually 0. Once the start bit is sent, the 8 bit data is sent serially, LSB (Least Significant Bit) is sent first and the MSB (Most Significant Bit) is sent last.

After the MSB is sent, there needs to be a message for the receiver to understand that the 8 bit data is sent. This end is indicated by the Stop bit. This is a 1, sent after the MSB is sent.

Parity Bit

This bit is additionally sent along with start, stop and data bits. This indicates the parity of sent bits. If the parity is odd, it is 1 and if the parity is even, the parity bit is 0. This bit usually is used as a further check in the correctness of the data.

Baud Rate

This is the data transfer rate, and is defined as the number of signal changes per second. This rate needs to be defined at both the transmitting and receiving end; else the communication will not take place.

RS232

This is the most widely used serial I/O interfacing standard among various devices and equipment. The voltage levels of RS232 are not TTL voltage levels, and hence, MAX232 voltage converter is used to be the interface between the two voltage levels.

Main.c

```
#include "main.h"

Void main(void)
{
    unsigned char SerialData;//to store data
    InitializeSerialCommunication();

    for(;;)                         //loop for ever
  {
            SerialData = ReceiveByteSerially();
        //Receive data from serial port and store to variable
            SendByteSerially(SerialData);
//Send a data to serial pot
    }
    //end of for loop
}
    //end of function
```

Main.h

```
#include    <htc.h>
#include "serial.h"

//DEFINE CONSTANTS
#define BAUD 9600
#define FOSC 16000000L

//SET CONFIGURATION BITS
__CONFIG (0x2F0A);
```

Serial.c

```
#include "serial.h"

void InitializeSerialCommunication(void)
{

    SPBRG = DIVIDER;        //LOAD BAUD RATE VALUE
    RCSTA = 0x90;           //ENABLE SERIAL PORT AND CONTINU
                              OUS RECEIVE
    TXSTA = 0x24;           //ENABLE TRANSMIT BIT AND HIGH
                              SPEED
}
```

```
void SendByteSerially (unsigned char byte)
{
    while(!TXIF)   //flag is set when transmit buffer is empty
    continue;
    TXREG = byte;
}

unsigned char ReceiveByteSerially()
 {
    while(!RCIF)   //flag is set when receive buffer is not empty
      continue;
    return RCREG;
}
```

Serial.h

```
#define DIVIDER ((int)(FOSC/(16UL * BAUD) -1))

    //FUNCTION PROTOTYPES
    unsigned char ReceiveByteSerially();
    void SendByteSerially (unsigned char byte);
    void InitializeSerialCommunication(void);
```

LCD Interfacing

LCD modules are commonly used today in projects related to microprocessors and microcontrollers. These devices are easy to code and act as a great device to display output real time basis. They have a definite number of display boxes. The 16 × 2 module is most commonly used (shown in figure).

The LCD module is readily available in market. It has 14 pins. The V_{SS} pin is ground, V_{CC} pin needs to be supplied with +5 volts power supply, and V_{EE} is used to give power supply to contrast.

The bit RS is used to select data register on setting it low, and is used to select command register on setting it high. The R/W is the read-write bit. Setting it enable reading and resetting it enables writing. E is the enable bit. The rest 8 bits DB0 to DB7 are the data bus lines used to send information to LCD or read contents of LCD.

The code to interface PIC microcontroller with the LCD is explained in the sections to come.

The starting address of the first line is 0x80 and the second position in the first line corresponds to 0x81 and so on. The starting address of the second line is 0xC0 and it goes on incrementing similar to the first line.

The LCD can be interfaced either in the 8 bit mode or in the 4 bit mode.

In the 8 bit mode the LCD requires 8 data lines and in the 4 bit mode it requires 4 data lines. Here we have used the 4 bit mode.

There are two methods of LCD interfacing namely The Busy flag method which involves the microcontroller sending commands to the LCD and the LCD completes those set of commands and then responds back to the microcontroller. The other method uses a delay approach which involves sending the required commands and waiting for a fixed duration of time and then sending the next command. The former is a system with feedback and the latter is a system without feedback. Here we have used the delay approach to avoid complexity.

Main.c

```
#include"main.h"
    void main(void)
    {
        unsigned char Count;
        MainSystemInitialize();
        for(;;)
        {
                MainDisplayName();
                MainDisplayMessage();
                Delay(2000);
        }
    }
    static void MainSystemInitialize(void)
    {
        LCDInitialize();
    }
    static void MainDisplayName(void)
    {
        LCDClear();
        LCDRow1();
        LCDWriteString("Electronics");
        Delay(500);
        LCDRow2();
        LCDWriteString("Simplified");
        Delay(3000);
    }
    static void MainDisplayMessage(void)
    {
        LCDClear();
        LCDRow1();
        LCDWriteString("PIC 16f877a");
        LCDRow2();
        LCDWriteString("HPS KIT V6.0");
    }
```

Main.h

```
#include <htc.h>
#include "delay.h"
#include "lcd.h"

#define _XTAL_FREQ  16000000
#define PRESSED                 0

//DEFINE CONSTANT
#define PRESENT                 1
#define DETECTED     1
#define NOT_DETECTED            0

//DEFINE MACROS
__CONFIG (0x2F0A);

//MACROS
//NO MACROS DEFINED

//FUNCTION PROTOTYPES
void main(void);
static void MainSystemInitialize(void);
static void MainDisplayName(void);
static void MainDisplayMessage(void);
```

LCD.c

```
#include "LCD.h"
    static void LCDEnable(void)
    {
     LCDEnablePin = HIGH;
    Delay(10);
    LCDEnablePin = LOW;
    }
    void LCDWriteCommand(LCDubyte LCDData)
    {
    LCDRegisterSelectPin = LOW;
    LCDWriteByte(LCDData);
    }
    void LCDWriteData(LCDubyte LCDData)
    {
      LCDRegisterSelectPin = HIGH;
     LCDWriteByte(LCDData);
    }
     void LCDWriteString(const char *lcd_string)
    {
          while(*lcd_string)
          {
                    LCDWriteData(*lcd_string++);
          }
    }
```

ADC Interfacing

ADC is the abbreviation for Analog to Digital Convertor. PIC series of microcontrollers have an inbuilt 10 bit ADC. ADCs are very important when it comes to data acquisition from field sensors.

The ADC converts the analog output from the sensor to digital form, and then we perform certain calculations to obtain back the analog value, and then accordingly display the required measurement of the physical quantity. The figure below shows the operation of ADC.

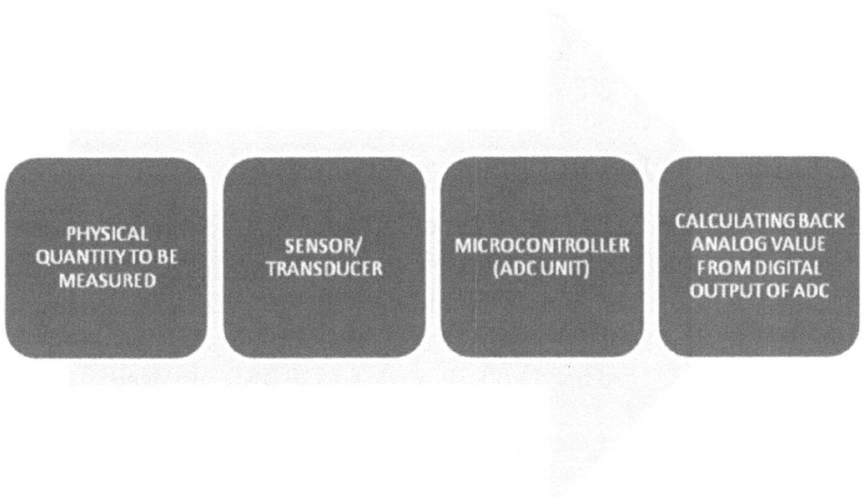

The results are usually preferred to be displayed on the LCD, or can be sent to a PC serially for further data processing or storage.

We now look into the various registers used for the ADC, namely **ADCON0 and ADCON1.**

We now look into the various bits under these two registers.

ADCS2:ADCS0

These bits are used to decide the clock conversion time. Depending on the frequency of the crystal oscillator used, care must be taken that the conversion time is greater than 1.6ms. This is a limitation of the controller.

Channel Selection Bits CHS2:CHS0

They are used to select the port A pin wherein you would be applying the analog input. Any port can be chosen from AN0 to AN7.

A/D Conversion Status Bit

If the bit is set, it means A/D conversion is in process. If it gets reset, it means the conversion process is over.

ADON or A/D on Bit

It this bit is set, the A/D feature is powered ON. In normal operation the A/D feature is kept OFF so that it does not consume power.

ADFM or A/D Result Format Selection Bit

The output of the ADC is a 10 bit output and is stored in two registers ADRESH and ADRESL. If the bit is set, the output is stored right justified, *i.e.,* 2 bits of ADRESH and all 8 bits of ADRESL are used to store the output. If the bit is made zero, the output is left justified, *i.e.,* 2 bits of ADRESL are used and all bits of ADRESH are used.

PCFGs or A/D Port Configuration Control Bits

These bits are used to mainly select which lines are going to acts as Analog inputs and which lines are digital inputs. Secondly, they are also used to instruct controller whether we are using a different value of Vref+ for our use.The steps adopted in performing the A/D process and the code is explained in the section of applications.

Main.c

```
#include"main.h"

void main(void)
{
     float temp;
     TRISE |= 0x07;
     PORTE |= 0x07;
     MainSystemInitialize();

  for(;;)
  {
               temp = ADCReadData(CHANNEL1);
               LCDWriteCommand(0x80);
                  LCDWriteFloat(temp);
               temp = ADCReadData(CHANNEL2);
               LCDWriteCommand(0x88);
 LCDWriteFloat(temp);
      temp = ADCReadData(CHANNEL3);
LCDWriteCommand(0xC0);
                  LCDWriteInteger(temp);
      temp = ADCReadData(CHANNEL4);
LCDWriteCommand(0xC8);
                  LCDWriteInteger(temp);
           temp = ADCReadData(CHANNEL5);
LCDWriteCommand(0x94);
                  LCDWriteInteger(temp);
```

```
           temp = ADCReadData(CHANNEL6);
LCDWriteCommand(0x9C);
                  LCDWriteInteger(temp);
      temp = ADCReadData(CHANNEL7);
               LCDWriteCommand(0xD4);
                  LCDWriteInteger(temp);
               temp = ADCReadData(CHANNEL8);
               LCDWriteCommand(0xDC);
                  LCDWriteInteger(temp);
}
}

static void MainSystemInitialize(void)
{
     ADCInitialize();
     LCDInitialize();
}
```

Main.h

```
#include <htc.h>
#include "delay.h"
#include "lcd.h"
#include "adc.h"

__CONFIG (0x2F0A);

void main(void);
static void MainSystemInitialize(void);
static void MainDisplayCompanyName(void);
static void MainDisplayMessage(void);
```

Adc.c

```
#include "adc.h"

void ADCInitialize(void)
{
    ADON = 1;                 //A/D converter module is powered up
    ADFM = 1      //Left justified.
 ADCS2 = ADCS1 = ADCS0 = 0;  //Fosc/2
  PCFG3 =PCFG2 = PCFG0 = 0;  PCFG1 = 1;
}

float ADCReadData(unsigned char ChannelNumber)
{
    unsigned char count, x,y,z;  float p=0,f;

    ADCON0 = ((ADCON0 & 0xc7) | (ChannelNumber));
    Delay(1);
    ADON = 1;
    ADCON0 |= 0x04;          //start conversion
    while( (ADCON0 & 0x04) == 0x04);
     //Wait for conversion to complete
    x=ADRESH;    //ADC result higher 2 bits
    y=ADRESL;    //ADC result lower 8 bits
    z=(x*256)+y;  //combined 10 bit number formation
    f=z;        //unsigned to float conversion
    p =(f*5)/1023;
   //equivalent voltage calculation from ADC output      Delay(1);
    return p;}
```

ADC.h

```
#ifndef ADC_H
#define ADC_H

#include <htc.h>
#include "delay.h"

typedef unsigned char ADCUbyte;

#define AVERAGEVALUE    5

#define CHANNEL1  0x00
#define CHANNEL2  0x08
#define CHANNEL3  0x10
#define CHANNEL4  0x18
#define CHANNEL5  0x20
#define CHANNEL6  0x28
#define CHANNEL7  0x30
#define CHANNEL8  0x38

void ADCInitialize(void);
float ADCReadData(unsigned char ChannelNumber);

#endif
```

Timers

When you want to generate a fixed time delay in executing certain instructions, or want to measure the time of events occurring outside the microcontroller, a special feature called timer is used. Depending on the use, there are two modes namely timer mode and counter mode.

In the timer mode, we use internal clock source, whereas in the counter mode, we input pulses from the outside using one of the pins of the microcontroller. In PIC16F877A, we have three timers—timer 0, timer 1 and timer 2.

Timer 0

Timer 0 is the most basic timer used. It is 8 bit in size and can count from 00h to FFh. The register that counts is termed as TMR0. There are a number of bits that need to be designated a value in order to use this timer the way we desire.

These are as follows:

T0CS or TMR0 Clock Source Select Bit

If the bit is high, the timer 0 module acts as a counter and if it is low, the timer 0 module acts as timer.

T0SE or TMR0 Source Edge Select Bit

This is used when timer 0 is used in counter mode. This basically defines as to when you want the counter to increment, on high to low transition or on low to high transition. Accordingly, the bit is set high or low respectively.

PSA or Prescalar Assignment Bit

This bit is used to define whether the prescalar is to be assigned to the timer 0 module or the watch dog timer.

PS2:PS0 Prescalar Rate Select Bits

These bits are used to define the rate of the timer. These three bits in all have 8 different rate select options and any one of them can be used by keeping the corresponding bit configuration.

Timer 1

This is a 16 bit timer/counter. It consists of two eight bit registers namely—TMR1H and TMR1L. The timer can count or increment from 0000h to ffffh.

We now understand the various bits that need to be defined for use of the timer 1 module.

T1CKPS1:T1CKPS0 or Timer 1 Input Clock Prescale Select Bits

These decide the pre-scalar value. There are four options and the two bits can be decided based on the option we choose.

T1OSCEN or Timer 1 Oscillator Enable Control Bit

This bit when set high enables the oscillator, else the bit is shut down.

TMR1CS or Timer 1 Clock Source Select Bit

This bit is kept low for timer operation, but is set high when we want to take external input to controller in counter mode.

TMR1ON or Timer 1 on Bit

Setting this bit high enables the timer whereas setting this bit low disables the timer.

The timer 1 operation and code is explained later on in the book.

Interrupts

This is a feature that is normally used along with timers. To understand interrupts, one needs to understand the difference between interrupts and polling. Hence, we first understand the concept based difference between the two.

Consider yourself to be the manager of a gas station and consider a person X as your assistant. You need to count the number of white cars entering your station. You will have two approaches in it.

The first one would be as follows—As soon as a car enters the station, you inform Mr. X to go and check whether it is a white car, and return to you with the answer. So each time a car comes you will ask your assistant to go and check.

The second approach is different—You inform your assistant to stay outside and keep checking all the cars that are entering the gas station, and as soon as a white car comes, he should come and inform you. You instruct him only once, and then he will keep checking for a white car.

The first approach is analogous to polling and the second one is analogous to interrupts. Which approach do you think is better?

This would vary from application to application, but in general, the interrupt approach is better. This saves efforts, as you can see that in first approach in the above example, you had to tell your assistant time and again to go and check.

The interrupt approach is like a parallel processor in the microcontroller. Once you initialize and instruct it, it keeps running even when the execution of data in occurring, and as soon as the interrupt condition is satisfied, the control or data flow is shifted to the required set of commands under the interrupt.

Interrupts are used along with timers. For the timer 0, the interrupt occurs on overflow of the TMR0 register from FFh to 00h. To enable timer 0 interrupt, we need to make TMR0IE bit high. When the interrupt occurs, the TMR0IF or the interrupt flag bit is set high by the controller.

Accordingly, the control is shifted to the interrupt function. If we wish to restart the interrupt, the flag bit needs to be made low and the enable bit must be made high again.

The same approach is used for timer 1, except that here the flag bit is TMR1IF and the enabling bit is TMR1IE.

The concept of interrupts is used in many applications, and the use as well as the code is explained in the further sections.

EEPROM

EEPROM stands for electrically erasable programmable ROM. PIC16F877A has 256 bytes of EEPROM memory. EECON1 is the control register used for the purpose of reading and writing into the PROM memory.

EEPROM is essential as you can store data into the controller which can be useful for real time monitoring, setting up of preset levels, testing, recording data, etc. We now look into the various bits present under the EECON1 register.

EEPGD or Program/Data EEPROM Selection Bit

The bit is used to decide whether to access the program memory or data memory.

WRER or EEPROM Error Flag Bit

This bit indicates whether the write operation is completed or terminated.

WREN or EEPROM Write Enable Bit

It bit is set high, writing to EEPROM is permitted. Else, it is inhibited.

WR or WRITE Control Bit

This bit is set high to initiate a write cycle. The bit is cleared once the write cycle to the EEPROM is complete. The bit cannot be cleared in software. The hardware, as mentioned earlier, clears the bit.

RD or READ Control Bit

Setting this bit high initiates reading of the EEPROM memory. This bit can only be set, not cleared in software, like the WR bit.

The steps to write to EEPROM data memory are:

1. Check WR to see if a write is in progress.
2. Write the address to EEADR.

3. Write the 8-bit data value to be programmed in the EEDATA register.
4. Clear the EEPGD bit to point to EEPROM data memory.
5. Set the WREN bit.
6. Disable interrupts (if at all they are enabled).
7. Execute the special five instruction sequence:
    - Write 55h to EECON2 in two steps (first write it to W and then to EECON2)
    - Write AAh to EECON2 in two steps (first write to W and then to EECON2)
    - Set the WR bit 8. Enable interrupts (if you using interrupts).
8. Clear the WREN bit to disable program operations.
9. At the completion of the write cycle, the WR bit is cleared and the EEIF interrupt flag bit is set.

Main.c

```
//INCLUDE MAIN HEADER FILE
#include"main.h"
unsigned char i,count1=0;
//MAIN PROGRAM BEGINS
void main(void)
{

    unsigned char Count,x;
    MainSystemInitialize();
    LCDDisplayInitializing();
    EEADR=0x00; //Specifying address to write to EEPROM
    EEDATA='c';   //Assigning the character c to be written in EE-
PROM
    EEPGD=0;
    WREN=1;
    GIE=0;
    EECON2=0x55;
    EECON2=0xAA;
    WR=1; //Writing to EEPROM memory
    GIE=1;
    WREN=0;
```

```
        LCDClear();
        LCDWriteString("character written");
        LCDRow1();
        LCDWriteData(EEDATA);
        Delay(3000);

        EEADR=0x00;  //Specifying address to read
        EEPGD=0;
        RD=1;  //Reading data from EEPROM
        LCDClear();
        Delay(2000);
        LCDClear();
        LCDWriteString("character read");
        LCDRow2();
        LCDWriteData(EEDATA);
    Delay(2000);
        }

    static void MainSystemInitialize(void)
    {
        LCDInitialize();
    }

    static void write(void)
    {
        count1=0;
        for(i=48;i<59;i++)
        {
        LCDClear();
        LCDWriteCommand(0x83);
        LCDWriteData(i);
        Delay(500);
        }
    }
```

Proximity Sensor Based Attendance System

The following block diagram shows how the system works. The proximity sensors A and B are located at the entry and exit of a room respectively.

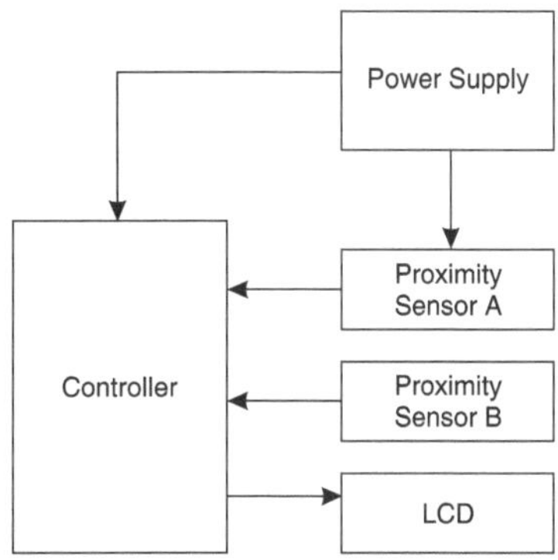

These sensors give a high output when an object comes in its vicinity. The output of the sensors is sent to the controller.

It processes and displays the number of people present in room on the display.

The following code shows how the attendance monitoring system works.

Main.c

```
#include"main.h"

main()
{
    Unsigned int count=0;
TRISB=1;      //Setting PORTB pins as input
    PORTB=0;          //Setting all pins of PORTB initially low
    RBPU=0;                   //enabling internal pull up resistors

    //RB0 is the input from proximity sensor A (At entry)
    //RB1 is the input from proximity sensor B (At exit)

    while(1)
    {
        if(RB0==1)
            {
            count++; //person has entered room
                    }

            if(RB1==1)
            {
            count--; //person has exit room
            }
    LCDWriteInteger(count);

    }
}
```

Main.h

```
#ifndef MAIN_H
#define MAIN_H

//INCLUDE HEADER FILE
#include <htc.h>
#include "delay.h"
#include "LCD.h"

__CONFIG(0x2F0A);

void TimerInitialize(void);

#endif
```

Count is essentially the variable that shows the number of people present in the room. We keep checking the pins RB0 and RB1. If RB0 turns high, we increment the count by 1, which indicates that a person has entered the room. If the pin RB1 turns high, we decrement count by 1, indicating that one person has left the room.

The count is displayed on the LCD. The checking of the pins goes on repeating, and we obtain the instantaneous value of the number of people present in the room.

MQ 2 Based Smoke/ Alcohol Detector

The following block diagram shows how the system operates. The gas sensor provides a digital output. This is fed to the controller.

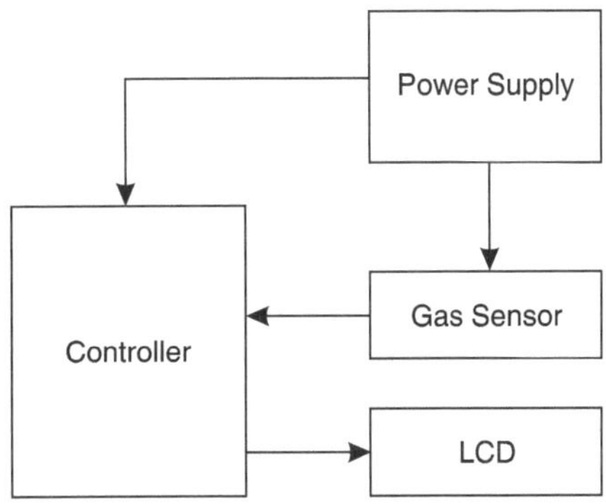

The controller continuously checks the output of the gas sensor, and if the output goes form low to high, it displays on the LCD that smoke/ alcohol is detected.

The figure shows how the gas sensor module looks like. The gas sensor works at elevated temperatures and hence, there is a coil placed internally to heat the module.

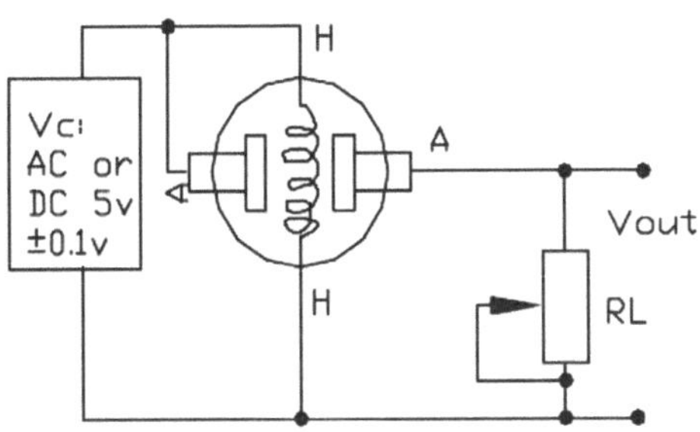

The module looks like this. The potentiometer is used to set the set point, and a comparator network compares the output with the set point. The output of the comparator is the digital signal that is given to the controller input pin. If the pin goes high, the controller displays on the LCD that smoke/alcohol is detected.

Main.c

```
#include"main.h"

main()
{
    TRISB=1;        //Setting PORTB pins as input
    PORTB=0;        //Setting all pins of PORTB initially low
    RBPU=0;         //enabling internal pull up resistors

//RB0 is the input from alcohol sensor RB1 is the input from smoke sensor

    while(1)
    {
            if(RB0==1)
            {       LCDRow1();
            LCDWriteString("Alcohol Detected");
            }
            else
            {   LCDRow1();
            LCDWriteString("Alcohol Absent");
            }

            if(RB1==1)
            {   LCDRow2();
            LCDWriteString("Smoke Detected");
            }
            else
            {   LCDRow2();
            LCDWriteString("Smoke Absent");
            }
}
}
```

Main.h

```
#ifndef MAIN_H
#define MAIN_H

//INCLUDE HEADER FILE
#include <htc.h>
#include "delay.h"
#include "LCD.h"

__CONFIG(0x2F0A);

void TimerInitialize(void);
#endif
```

The code shows the various steps to accept input and display the results. The PORTB is used as the input port and pins RB0 and RB1 are used as the inputs to the controller. One must remember to enable to enable the pull up resistors. Next, we check if the pins are high or low. If the pin goes high, we say that the particular compound is present. Else, we say it is absent. We use two modules of sensor, one for smoke and the other for alcohol.

Note that one needs to include the LCD files in the project for the program to work.

LED Brightness Control

LED brightness control is also based on the concept of pulse width modulated wave. The effective time, the LED is on, hence, controlling its brightness. If the pulse has more time of logic high and lesser time of logic low, the LED will glow brightly, else dimly. We use timers to generate the required pulse.

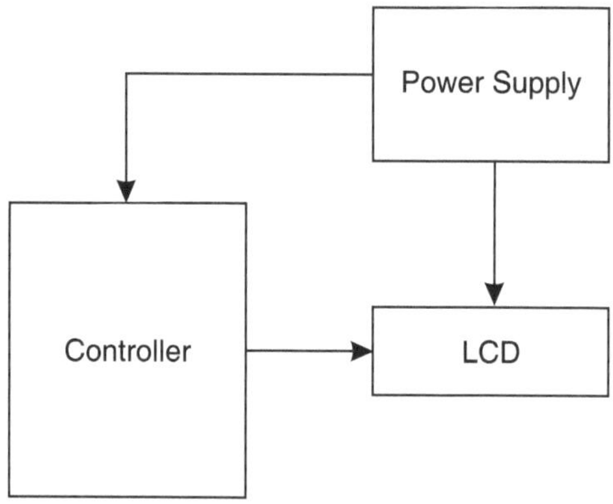

Main.c

```
#include "main.h"

    unsigned int stepsize=0,count=0, cycles=0;

    interrupt timer(void)
    {
        if(TMR1IF==1)
        {
                if(count>=(115-stepsize))
                {
                        if(count==115)
                        {       count=0;
                                cycles++;
                        }
                        RB1=0;
                }

                else if(count<(115-stepsize))
                {       RB1=1;      }

                if(cycles%50==0)
                        {
                if(stepsize>=115)
                                {stepsize=0;}
                              else
                          stepsize=stepsize+3;
                        }
                count=count++;
                TMR1IF=0;
                TMR1IE=1;
                TMR1H=0xfe;
        }
    }
```

```
void main(void)
{
    float temp=0;
    int i;

    TRISB=0x00;
    PORTB|=0xA0;
    ADCInitialize();
    LCDInitialize();
    LCDClear();

    TimerInitialize();

    while(1);
}

void TimerInitialize(void)
{
    TMR1H=0xfe;
    T1CKPS1=0;     //setting prescalar 1:1
    T1CKPS0=0;
    TMR1ON=1;
    GIE=1;
    PEIE=1;
    TMR1IE=1;
}
```

Main.h

```
#ifndef MAIN_H
#define MAIN_H

//INCLUDE HEADER FILE
#include <htc.h>
#include "delay.h"
#include "LCD.h"

__CONFIG(0x2F0A);

void TimerInitialize(void);

#endif
```

The above code shows how exactly we control the LED brightness. The LED is connected at the pin RB1. We set it high for the ON time and set it low for the OFF time.

We first understand the various variables used. The variable "count" is the time taken by the timer to reach 0xff. It is used to generate the logic high and logic low pulses to form one wave or cycle. The variable "cycles" is used to count the number of cycles that are passed in the wave. This is important to change the ON and OFF times, as it controls the increment of the variable step size. After a certain number of cycles, increment the variable "stepsize", which alters the ON and OFF times, thus, changing the effective Led brightness.

In the following code, we change the step size by 3 for every 50 cycles passed. These numbers are adjusted to obtain a gradual but distinct change in LED brightness.

Speed Control of DC Motor

In this section, we are controlling the speed of the DC motor using Pulse Width Modulated Waveform. Refer the following block diagram.

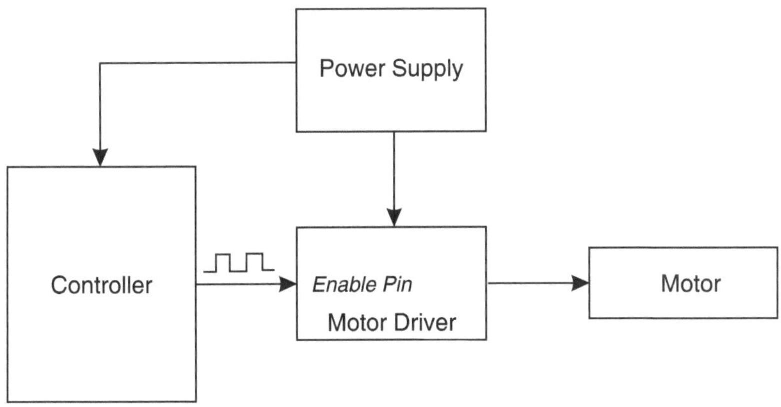

The controller generates a square waveform (PWM). This waveform is sent to the enable pin of the motor driver IC. You must recollect that the enable pin of the motor driver unit enables or disables the motor. If the pin is set low, the motor stops, and if the pin is set high, the motor motion is enabled. This enabling and disabling is done by the waveform generated by the controller.

If the waveform has a higher logic 1 time and a lesser logic 0 time, the motor motion is faster/RPM is higher. Else, if the logic high is for a lesser time and the logic low is for a higher time, the speed of the motor is low.

The following code shows how we generate a pulse width modulated wave, and how we control the speed.

Main.h

```
#ifndef MAIN_H
#define MAIN_H

//INCLUDE HEADER FILE
#include <htc.h>
#include "delay.h"
#include "LCD.h"

__CONFIG(0x2F0A);

void TimerInitialize(void);

#endif
```

Main.c

```
#include "main.h"

unsigned int temp2=8,count=0, count1=0;

interrupt timer(void)
{
    if(TMR1IF==1)
    {
            if(count<(115-temp2))
            {
                    RB0=1;
            }
            else if(count>=(115-temp2))
            {
                    if(count==115)
                    {
                            count=0;
                            count1++;
                    }
                    RB0=0;
            }
    if(count1%150==0)
                    {
                    if(temp2>=115)
                    {temp2=0;}
                    else
                    temp2=temp2+3;
                    }
            count=count++;
            TMR1IF=0;
            TMR1IE=1;
            TMR1H=0xfe;
    }
}
```

```
void main(void)
{
    float temp=0;
    int i;

    TRISB=0x00;
    PORTB|=0xA0;
    ADCInitialize();
    LCDInitialize();
    LCDClear();
    TimerInitialize();

    for(;;);
}

void TimerInitialize(void)
{
    TMR1H=0xfe;
    T1CKPS1=0;      //setting prescalar 1:1
    T1CKPS0=0;
    TMR1ON=1;
    GIE=1;
    PEIE=1;
    TMR1IE=1;
}
```

We have used timers to generate the wave. We set it to a particular frequency by making necessary calculations (learnt in the section named Timers). Next, we keep the entire wave time in counts. We have hence named the corresponding variable Count. The time of one count is kept as the time for the timer 1 to overflow.

We need to demarcate the time for which logic high is sent and for how much time logic low is sent. This is decided by the variable temp2. This variable basically demarcates the counts as high or low.

Temp2 is varied used another variable count1. Count 1 is incremented by 1 about every cycle of the wave generated. The variable temp2 is increased by 3 after count1 is incremented every 200 times.

We thus can see the gradual change in the speed of the dc motor.

Temperature Controlled Fan

The following block diagram shows as to how we use a LM35 temperature sensor to control the switching ON-OFF of a fan.

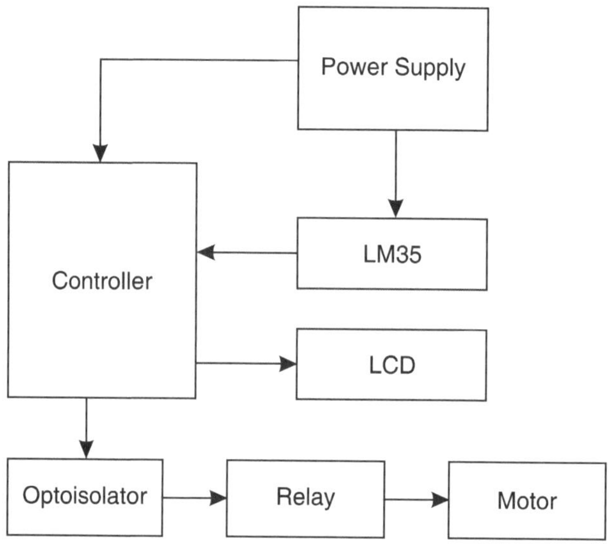

The LM35 sensor provides an analog output which is fed to the ADC pin of the controller. The controller converts it into the required digital value, then, mathematically getting back the analog value and displaying it on the screen.

Also, we need to control the fan. So it sets a required condition that if the temperature goes above 30 degree Celsius, the fan must be put ON. So, the controller generates the required control signal which is sent to the optoisolator.

Optoisolator is used to ensure that excessive currents if induced in the output side do not damage the controller. The output of the optoisolator is then given to the relay (5 V input actuation voltage) control coil. The relay output is connected to the motor.

So when the temperature exceeds the set point, the controller sends the control signal via the optoisolator, which actuates the relay, and in-turn the relay switches ON the fan.

This system can be used for a lot of applications, one of which includes Cooling Fan Pads for laptops.

Main.c

```
#include"main.h"

void main(void)
{
    float temp;
    TRISB = 0x00;
    TRISE |= 0x07;
    PORTE |= 0x07;
    MainSystemInitialize();

for(;;)
    {
                temp = ADCReadData(CHANNEL1);
                temp=temp*100;
                LCDWriteCommand(0x80);
                LCDWriteFloat(temp);

                if(temp>35)
                        RB1=1;
                else if(temp<35)
                        RB1=0;
    }
}

static void MainSystemInitialize(void)
{
    ADCInitialize();
    LCDInitialize();
}
```

Adc.c

```
#include "adc.h"

void ADCInitialize(void)
{
    ADON = 1;               //A/D converter module is powered up
    ADFM = 1;               //Left justified
ADCS2 = ADCS1 = ADCS0 = 0;                      //FOSC/2
 PCFG3 = PCFG2 = PCFG0 =0;PCFG1 = 1;
}

float ADCReadData(unsigned char ChannelNumber)
{
    unsigned char count,I, x,y,z;
    float p=0,f;
    ADCON0 = ((ADCON0 & 0xc7) | (ChannelNumber));

for(i=0;i<100;i++)
{
    ADON = 1;
    ADCON0 |= 0x04;                     //start conversion
    while( (ADCON0 & 0x04) == 0x04);
    //wait for conversion to complete

    x=ADRESH;  //ADC result higher 2 bits
    y=ADRESL;  //ADC result lower 8 bits
    z=(x*256)+y;   //combined 10 bit number formation
    f=z;          //unsigned to float conversion
    p +=(f*5)/1023;  //equivalent voltage calculation from ADC out-
put
    Delay(1);
}
p/=100;
    return p;
}
```

Adc.h

```
#ifndef ADC_H
#define ADC_H

#include "delay.h"

//USER DEFINED DATA TYPE
typedef unsigned char ADCUbyte;

//DEFINE CONSTANT
#define AVERAGEVALUE 5
#define CHANNEL1 0x00
#define CHANNEL2 0x08
#define CHANNEL3 0x10
#define CHANNEL4 0x18
#define CHANNEL5 0x20
#define CHANNEL6 0x28
#define CHANNEL7 0x30
#define CHANNEL8 0x38

//FUNCTION PROTOTYPE
void ADCInitialize(void);
float ADCReadData(unsigned char ChannelNumber);

#endif
```

Main.h

```
#include <htc.h>
#include "delay.h"
#include "lcd.h"
#include "adc.h"
#include "serial.h"

//DEFINE MACROS
__CONFIG (0x2F0A);

//FUNCTION PROTOTYPES
void main(void);
static void MainSystemInitialize(void);
static void MainDisplayCompanyName(void);
static void MainDisplayMessage(void);
```

The above two codes are the ones used to perform the particular application. In addition, you need to include the LCD codes for proper functioning.

The code continuously checks for the temperature. The ADC code gives us the required temperature value. It is sent back to the main.c code. Here, it is compared with 35, if it exceeds 35, the signal is sent to switch ON the fan, else the fan remains in OFF condition.

Ultrasonic Sensor Based Distance Measurement

This project deals with measuring distance with the help of an ultrasonic sensor. The block diagram shown below explains the entire mechanism of the system.

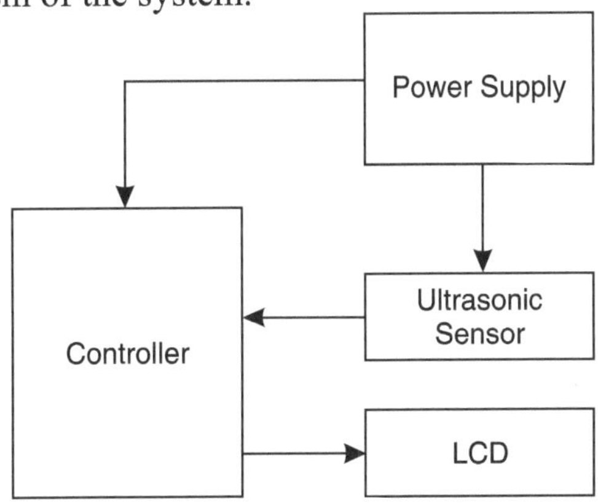

The ultrasonic sensor sends gives an output which is received by the controller. The controller uses timers for the operation. The timer measures the time taken by the wave to return and scales it, to obtain the distance in centimeters.

The controller then sends the data to the LCD to be displayed.

Main.c

```
#include<htc.h>
    #include "LCD.h"
    #include "delay.h"

    unsigned int TIME,TOTAL;

    interrupt timer(void)
    {
        if(TMR0IF == 1)
```

```
    {
            TIME += 256;
    }

 TMR0IF=0;
     TMR0IE=1;
}

void main(void)

{
  TRISB |= 0xFF;
  TRISC &= 0x00;
  LCDInitialize();
  LCDWriteString("DISTANCE : ");

     while(1)
{
     RC4 = 0;
     RBPU = 0;
     TOTAL =0;
     TIME =0;
   TMR1H  = 0xFF;

 TMR1L  = 0xF6;
T1CON  = 0x20;
     TMR1IF=0;
     RC4=1;
     TMR1ON = 1;

     while(!TMR1IF);
     RC4 = 0;

  RBPU = 1;
  while(!RB0);
  TMR0=0;
  T0CS = 0;
  T0SE=0 ;
  PSA = 0;
  PS2=PS1=PS0=1;
  GIE = 1;
  TMR0IE= 1;
```

```
while(RB0);
TOCS=1;
GIE = 0;
 TOTAL = TIME+TMR0;
if(TOTAL<=256)
 {LCDDisplayByte(0x8B,TOTAL);}
else
 {LCDDisplayByte(0x8B,256);}

Delay(200);
 }
}
```

The above code uses timers 0 and 1. The two timers are used for two different purposes. One is to generate a 38 kHz trigger pulse, and the other is used to count the time the ultrasonic wave (emitted by the transmitter) takes to return.

For the timer 1, the registers are given the required values, so that the equivalent time corresponds to a single pulse of frequency 38 kHz. The pin RC4 is set high for the required time, generating the pulse, and this pin is connected to the trigger input of the ultrasonic sensor module.

Once the trigger is applied, the module emits an ultrasonic wave. Now we start the other timer, timer 0. This timer keeps on counting till the pin RB0 turns low. The mechanism of the module is that once the module receives back the wave, the echo pin, which is initially high, turns low. As soon as this pin turns low, we stop the timer 0. The distance in cm is the value of the TMRO register.

Note that the maximum value TMR0 can hold is 0xff or 256. Hence, it cannot measure a distance greater than 256. To enable the system to measure distance more than 256 cm, we state that if the interrupt flag bit is set high for timer 0, we add additional 256 to the variable TIME, and restart the process of counting immediately till the pin RB0 turns low. We add 256 and the new TMR0 value, to obtain the final distance.

One must note that the maximum range of the ultrasonic sensor is about 300 cm. one cannot measure distances beyond this using this device.

The files LCD.c and LCD.h must be included for the code to work properly.

High Speed Multiple Sensor Based Data Acquisition System

A data acquisition system is used to collect data from any system for future analysis or for monitoring one or more parameters for various reasons ranging from calculation of system efficiency to generation of log files for statistical data.

Ideally any data acquisition system must possess a number of channels and must obtain the data from the channels in parallel so that even minute changes can be detected. But generally such systems are very expensive.

This system establishes a low cost and high speed mechanism to develop multiple channel Data-Acquisition system capable of acquiring both Analog or Digital data from various sensors monitoring various physical parameters namely light intensity, Temperature, Magnetic property of a material, presence of various gases, relative humidity, proximity, Distance and differential pressure.

Although sensing systems are readily available in the embedded market but they cater only to sensing or measurement of a single parameter. However in most industries and processing plants a large number of parameters are to be monitored, hence they procure a number of sensing systems to cater to different parameters. This makes troubleshooting and fault diagnosis extremely difficult and a time consuming process. A very few of these systems have inbuilt integration with GSM modem or PC for data acquisition, as a result of which in some cases separate PC and GSM modem interfaces are made and this adds to the complexity of the system as a number of functional units

A 12 channel data acquisition system is implemented using the PIC18f458 and PIC16f877a which are Harvard architecture based micro-controller. The system has three controllers performing dedicated tasks. The system is interfaced to a PC, 20x4 LCD display and a GSM modem.

The system block diagram clearly elucidates the working of the system developed

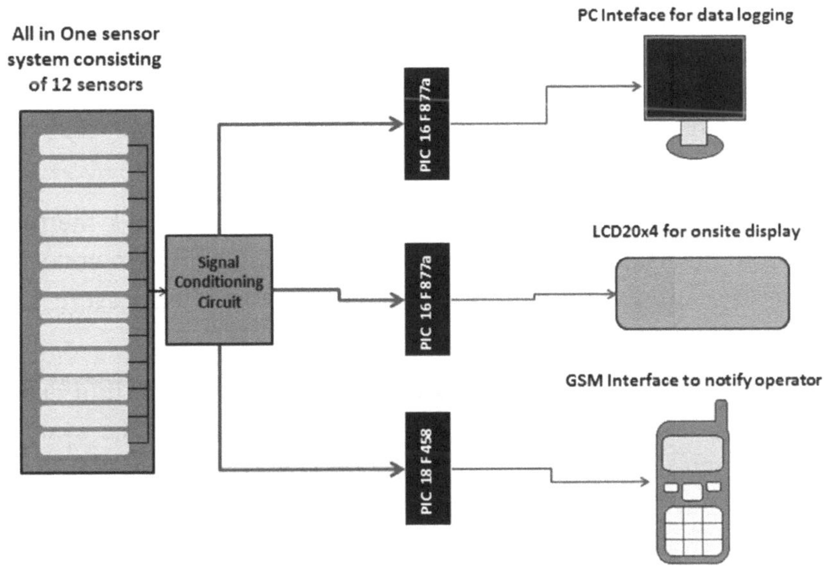

For analysis purpose the 12 channels are connected to 12 different sensors which include the following:

1. LDR for measuring luminous intensity
2. LM35 for measuring temperature
3. Thermistor for measuring temperature
4. IR proximity sensor
5. HI300 for measuring magnetic field strength
6. MQ2 used for monitoring presence of flammable gases
7. MQ3 used for monitoring presence of LPG
8. MQ4 used for monitoring presence of smoke
9. MQ6 used for monitoring presence of Butane etc.
10. SYHS202 for measuring humidity
11. MXP10DP for measuring differential pressure
12. HY202 ultrasonic eye for measuring distance precisely.

The Resistance of an LDR decreases almost linearly with linear decrease in luminous intensity. Hence the voltage divider arrangement is used and a time varying voltage is obtained depending on the luminous intensity. The two capacitors in parallel ensure a smooth response of the system when light intensity changes.

The LM35 is a precision integrated circuit temperature sensor whose output is directly calibrated in degree Celcius. The output voltage of the IC changes linearly with change in temperature. The LM35 has a full scale range of –40 to 150 degrees.

A Thermistor has a negative temperature coefficient. A 10 k thermistor is used to measure temperature and the voltage divider concept is used to measure the output voltage which in turn is a measure of the temperature of the system.

In order to measure the proximity of an object the IR proximity sensor is used. It consists of a IR led and an IR photodiode. The system is arranged such that the IR LED and IR photodiode are parallel to each other. Whenever an object comes in front of the system, the object reflects the IR rays emanating from the IR LED and it is received on the photodiode. The output of the photodiode is fed to the comparator which gives a bi-state output.

The HI300 is a miniature hall effect sensor where the output is the hall voltage. Whenever a magnetic field is encountered by the hall effect IC the HI300 outputs a hall voltage proportional to the magnetic field

The MQ series of gas sensors are small gas sensors composed of micro Al_2O_3 ceramic tubes, a SnO_2 as sensitive layer, measuring electrode and a heater fixed into a crust made by plastic and stainless steel net. The heater provides necessary work conditions for work of sensitive components. The enveloped MQ-4 have 6 pin, 4 of them are used to fetch signals, and other 2 are used for providing heating current. The differential output of the system is fed to a differential amplifier and the ppm concentration of the sample gas can be estimated after suitable calibration.

The SYHS 202 is a temperature compensated sensor which converts relative humidity to output voltage. The system has a linear response.

The MPX10 and MPXV10GC series devices are silicon piezo-resistive pressure sensors providing a very accurate and linear voltage output—directly proportional to the applied pressure.

The ultrasonic module based ultrasonic eye is used for distance measurement. It works on the echo principle. A 38 KHz Pulse is sent to the transmitter and the Ultrasonic transmitter send the ultrasonic beam after striking the object within range the Receiver gets the reflected ultrasonic beam. Upon knowing the time period between the transmission and reception of the beam the distance can be estimated precisely.

Adc.c

```
#include "adc.h"

void ADCInitialize(void)
{
    ADON = 1;                //A/D converter module is powered up
    ADFM = 1; //Left justified
 ADCS2 = ADCS1 = ADCS0 = 0;                    //FOSC/2
  PCFG3 = PCFG2 = PCFG0 =0;  PCFG1 = 1;
}

float ADCReadData(unsigned char ChannelNumber)
{
    unsigned char count;
    unsigned x,y,c=0xff,z;
    float p=0,f;

    ADCON0 = ((ADCON0 & 0xc7) | (ChannelNumber));
    Delay(1);
    ADON = 1;
    ADCON0 |= 0x04;                              // start conver-
sion

    while( (ADCON0 & 0x04) == 1); //wait for conversion to com-
plete

    x=ADRESH;  //ADC result higher 2 bits
    y=ADRESL;  //ADC result lower 8 bits
    z=(x*c)+y;   //combined 10 bit number formation
    f=z;       //unsigned to float conversion
    p =(f*5)/1023;  //equivalent voltage calculation from ADC output

 return p;
 }
```

Main.c

```
#include"main.h"

    void main(void)
    {
        float temp;
        unsigned int temp1;
        TRISE |= 0x07;
        PORTE |= 0x07;
        MainSystemInitialize();

    for(;;)
    {
                SendStringSerially("TEMPERATURE :");
                temp = ADCReadData(CHANNEL3);
                temp=(temp/2)*100;
                SendFloatSerially(temp);
                SendStringSerially(" C   ");
    Delay(200);
                temp = ADCReadData(CHANNEL5);
                if(temp>2.8)
                SendStringSerially("GAS DETECTED   ");
                else if(temp<2.8)
                SendStringSerially("GAS ABSENT   ");
    Delay(200);
                SendStringSerially("LUMINOUS INTENSITY :");
                temp = ADCReadData(CHANNEL2);
                temp*=2;
                temp1=temp;
                SendIntegerSerially(temp1);
                SendStringSerially(" Cd   ");
```

Serial.c

```
#include    "serial.h"
unsigned char flag1;

void InitializeSerialCommunication(void)
{
    SPBRG = DIVIDER;        // Load baud rate value
    RCSTA = 0x90; // Enable serial port and continuous receive
    TXSTA = 0x24;  // Enable transmit bit and high speed
    SYNC=0;
}

void SendByteSerially (unsigned char byte)
{   /* output one byte */
    while(!TXIF)      /* set when register is empty */
            continue;
    TXREG = byte;
}

unsigned char ReceiveByteSerially()
{   /* retrieve one byte */
    while(!RCIF)      /* set when register is not empty */
            continue;
    return RCREG;
}

void SendStringSerially(const char *serial_string)
{   while(*serial_string)
    {
    SendByteSerially(*serial_string);
    serial_string++;
}
}
void SendCharSerially(unsigned char Value)
{
    if(( Value / 100 ) == 0)
            SendByteSerially(0x20);                    //put space
    else
            SendByteSerially((Value / 100) | 0x30);
    if(((Value % 100) / 10) == 0)
            if(( Value / 100 ) == 0)
                    SendByteSerially(0x20);
            else
```

```
                SendByteSerially(((Value % 100) / 10) | 0x30);
    else
            SendByteSerially(((Value % 100) / 10) | 0x30);

            SendByteSerially((Value % 10) | 0x30);
}

void SendIntegerSerially(int IntegerNumber)
{
    unsigned char c[6];
    unsigned char i = 0;

    if(IntegerNumber < 0)
    {
            SendByteSerially('-');
            IntegerNumber = - (IntegerNumber);
    }

    if(IntegerNumber == 0)
            SendByteSerially(0x30);
    while( IntegerNumber > 0)
    {
            c[i++] = ((IntegerNumber % 10) | 0x30);
            IntegerNumber /= 10;
    }

    while(i-- > 0)
            SendByteSerially(c[i]);
}

void SendFloatSerially(float FloatNumber)
{
    unsigned char c[4];
    unsigned char i = 0;
    int n, Temp;

    if(FloatNumber < 0)
    {
            SendByteSerially('-');
            FloatNumber = -(FloatNumber);
    }
```

```
        n = FloatNumber;
        Temp = n;
        SendIntegerSerially(n);
        SendByteSerially('.');
        FloatNumber = ((FloatNumber - n) + 1) * 1000;
        n = FloatNumber;

        while(n > 0)
        {               c[i++] = (n % 10) + 48;
                n /= 10;
        }
i -= 1;
        while(i-- > 1)
                SendByteSerially(c[i]);
        if(Temp < 10)
        {
                SendByteSerially(" ");
        }
        if(Temp >= 10 && Temp < 100)
        {
                SendByteSerially(" ");
        }
    }
```

ADC.h

```
#ifndef ADC_H
#define ADC_H

#include "delay.h"
   typedef unsigned char ADCUbyte;

      #define AVERAGEVALUE          5
      #define CHANNEL1 0x00
      #define CHANNEL2 0x08
      #define CHANNEL3 0x10
      #define CHANNEL4 0x18
      #define CHANNEL5 0x20
      #define CHANNEL6 0x28
      #define CHANNEL7 0x30
      #define CHANNEL8 0x38

      void ADCInitialize(void);
      float ADCReadData(unsigned char ChannelNumber);
#endif
```

Main.h

```
#include <htc.h>
#include "delay.h"
#include "serial.h"
#include "adc.h"

//DEFINE MACROS
__CONFIG (0x2F0A);

void main(void);
static void MainSystemInitialize(void);
static void MainDisplayCompanyName(void);
static void MainDisplayMessage(void);
```

Serial.h

```
#include     <htc.h>          //INCLUDE PIC HEADER FILE

#define BAUD 9600
#define FOSC 16000000L

#define DIVIDER ((int)(FOSC/(16UL * BAUD) -1))

//FUNCTION PROTOTYPES
unsigned char ReceiveByteSerially();
void SendByteSerially (unsigned char byte);
void InitializeSerialCommunication(void);
void SendStringSerially(const char *serial_string);
void SendCharSerially(unsigned char value);
void SendIntegerSerially(int IntegerNumber);
void SendFloatSerially(float FloatNumber);
```